BEEF PRODUCTION FROM DIFFERENT DAIRY BREEDS AND DAIRY BEEF CROSSES

CURRENT TOPICS IN VETERINARY MEDICINE AND ANIMAL SCIENCE

VOLUME 21

Other titles in this series

Volume 1: Control of Reproduction in the Cow, edited by J.M. Sreenan

Volume 2: Patterns of Growth and Development in Cattle, edited by H. de Boer and J. Martin

Volume 3: Respiratory Diseases in Cattle, edited by W.B. Martin

Volume 4: Calving Problems and Early Viability of the Calf, edited by B. Hoffmann, I.L. Mason and J. Schmidt

Volume 5: The Future of Beef Production in the European Community, edited by J.C. Bowman and P. Susmel

Volume 6: Diseases of Cattle in the Tropics: Economic and Zoonotic Relevance, edited by M. Ristic and I. McIntyre

Volume 7: Control of Reproductive Functions in Domestic Animals, edited by W. Jochle and D.R. Lamond

Volume 8: The Laying Hen and its Environment, edited by R. Moss

Volume 9: Epidemiology and Control of Nematodiasis in Cattle, edited by P. Nansen, R.J. Jørgensen and E.J.L. Soulsby

Volume 10: The Problem of Dark-Cutting in Beef, edited by D.E. Hood and P.V. Tarrant

Volume 11: The Welfare of Pigs, edited by W. Sybesma

Volume 12: The Mucosal Immune System, edited by F.J. Bourne

Volume 13: Laboratory Diagnosis in Neonatal Calf and Pig Diarrhoea, edited by P.W. de Leeuw and P.A.M. Guinée

Volume 14: Advances in the Control of Theileriosis, edited by A.D. Irvin, M.P. Cunningham and A.S. Young

Volume 15: Fourth International Symposium on Bovine Leukosis, edited by O.C. Straub

Volume 16: Muscle Hypertrophy of Genetic Origin and its Use to Improve Beef Production, edited by J.W.B. King and F. Ménissier

Volume 17: Aujeszky's Disease, edited by G. Wittmann and S.A. Hall

Volume 18: Transport of Animals Intended for Breeding, Production and Slaughter, edited by R. Moss

Volume 19: Welfare and Husbandry of Calves, edited by J.P. Signoret

Volume 20: Factors Influencing Fertility in the Postpartum Cow, edited by H. Karg and E. Schallenberger

Series ISBN: 90-247-2429-5

BEEF PRODUCTION FROM DIFFERENT DAIRY BREEDS AND DAIRY BEEF CROSSES

A Seminar in the CEC Programme of Coordination of Research on Beef Production held at Castleknock, Co. Dublin, Ireland, April 13-15, 1981

Sponsored by the Commission of the European Communities, Directorate-General for Agriculture, Coordination of Agricultural Research

edited by

G.J. More O'Ferrall
The Agricultural Institute
Dunsinea Research Centre
Castleknock, Co. Dublin
Ireland

1982
SPRINGER-SCIENCE+BUSINESS MEDIA, B.V.

Library of Congress Cataloging in Publication Data

Main entry under title:

Beef production from different dairy breeds and
dairy beef crosses.

(Current topics in veterinary medicine and
animal science ; v. 21)
1. Beef cattle--Breeding--Congresses. 2. Dairy
cattle breeds--Congresses. 3. Beef cattle breeds
--Congresses. 4. Dual-purpose cattle--Congresses.
I. More O'Ferrall, G. J. II. Commission of the
European Communities. Coordination of Agricultural
Research. III. Series.
SF207.B468 1982 636.2'13 82-14480
ISBN 978-90-481-8275-6 ISBN 978-94-017-0847-0 (eBook)
DOI 10.1007/978-94-017-0847-0

Publication arranged by:
Commission of the European Communities,
Directorate-General Information Market and Innovation

EUR 8073 EN

CONTENTS

PREFACE IX

OBJECTIVES AND BACKGROUND OF SEMINAR X

OPENING OF SEMINAR XI
Mr. L. Allen, T.D., Minister of State for Agriculture

SESSION 1 : COMPARISON OF DIFFERENT STRAINS OF HOLSTEIN AND FRIESIAN AND OTHER DAIRY BREEDS FOR BEEF AND VEAL PRODUCTION

TESTING OF DIFFERENT STRAINS OF FRIESIAN CATTLE IN POLAND. BEEF PERFORMANCE OF F_1 FRIESIAN BULLS FATTENED UNDER INTENSIVE FEEDING CONDITIONS TO 450 kg or 550 kg LIVEWEIGHT.

L. Reklewski 1

Discussion. 20

COMPARISON OF BEEF PERFORMANCE OF DIFFERENT FRIESIAN STRAINS FROM PROGENY TEST RESULTS IN IRELAND.

J. Flanagan 21

Discussion 22

DANISH BREEDING STRATEGIES FOR MILK AND BEEF TRAITS IN DAIRY AND DUAL PURPOSE CATTLE BREEDS.

B. Bech Andersen 29

A NOTE ON THE COMPARISON OF HOLSTEINS AND FRIESIANS FOR GROWTH, FEED EFFICIENCY AND CARCASS TRAITS.

J.R. Southgate 37

Discussion 42

MEAT PRODUCTION OF HOLSTEIN FRIESIANS IN COMPARISON TO DUTCH FRIESIANS AND DUTCH RED AND WHITE.

J.K. Oldenbroek 45

ABILITY OF HOLSTEIN FRIESIAN, DUTCH BLACK AND WHITE, AND BELGIAN WHITE-RED BULL CALVES FOR VEAL AND BEEF PRODUCTION.

D.L. de Brabander, Ch.V. Boucqué, M. Casteels,
J.V. Aerts and F.X. Buysse 53

Discussion 66

Concluding comments on Session 1
D. Smidt 68

SESSION II : USE OF BEEF BREEDS FOR CROSSING IN DAIRY HERDS

RELATIVE PERFORMANCE OF BEEF BREEDS AND THEIR DAIRY CROSSES IN IRELAND.

T.T. Teehan.. 69

Discussion 81

SIRE BREED INFLUENCE OF VARIOUS BEEF BREEDS ON CALVING PERFORMANCE, GROWTH RATE, FEED EFFICIENCY, CARCASS AND MEAT QUALITY.

T. Liboriussen 82

Discussion 92

COMPARISON OF DIFFERENT SIRE BREEDS CROSSED WITH FRIESIAN COWS: PRELIMINARY RESULTS.

F. Menissier, J. Sapa, J.L. Foulley, J. Frebling and B. Bonaiti 94

BEEF PRODUCTION IN DANISH JERSEY. CHAROLAIS x JERSEY CROSSINGS.

A. Neimann-Sørensen and E.O. Neilsen 137

Discussion 150

PERFORMANCE OF BELGIAN WHITE-RED CROSSBREDS USED FOR BEEF PRODUCTION - PRELIMINARY RESULTS.

L.O. Fiems, Ch.V. Boucqué and F.X. Buysse 151

CROSSBREEDING OF BEEF BULLS ON FRIESIAN COWS IN ITALY.

A.L. Catalano.. 161

Discussion... 171

FATTENING PERFORMANCE AND SLAUGHTER TRAITS OF PUREBRED AND CROSSED YOUNG BULLS.

M. Bonsembiante, G. Bittante, P. Cesselli and I. Andrighetto 172

THE EFFECT OF FREEZING ON THE MEAT COLOUR OF PUREBRED AND CROSSED YOUNG BULLS.

G. Bittante and I. Andrighetto 188

EARLY CALVING OF CROSSBRED HEIFERS. I COMPARISON BETWEEN ANIMALS SLAUGHTERED JUST AFTER AND SEVEN MONTHS AFTER FIRST CALVING.

A. Romita, S. Gigli and A. Borghese 202

Discussion 219

SELECTION OF BEEF BULLS FOR TERMINAL CROSSING IN FRANCE.

J.L. Foulley and F. Menissier........................... 221

Discussion.. 244

Concluding comments on Session II
R.B. Thiessen... 246

SESSION III : MODELLING STUDIES ON THE BALANCE BETWEEN MILK AND BEEF

INFLUENCE OF ECONOMIC WEIGHTS AND POPULATION STRUCTURE ON SELECTION RESPONSE OF MILK AND BEEF TRAITS.

D. Fewson 247

Discussion 257

PROGENY TESTING OF DAIRY BULLS FOR MEAT PRODUCTION AND AGGREGATION WITH DAIRY INDICES

J.J. Colleau 258

Discussion 266

COMPARISON OF DANISH FRIESIAN, FRIESIAN AND RED DANISH FOR BEEF PRODUCTION

B. Bech Andersen 268

Discussion.. 277

ECONOMIC AND GENETIC OPTIMISATION OF DUAL PURPOSE BULL TESTING AND SELECTION

E.P. Cunningham and B. Moioli........................... 279

Discussion 293

THE ECONOMIC COMPARISON BETWEEN THE HOLSTEIN AND NORMAN BREEDS

J.J. Colleau and G. Lienard............................. 295

Discussion 319

Concluding comments on Session III
A. Neimann-Sorensen 321

SESSION IV : PRESENT PRACTICE AND FUTURE POSSIBILITIES FOR INCREASING BEEF PRODUCTION FROM DAIRY HERDS

USE OF BEEF BREEDS FOR CROSSING ON DAIRY HERDS IN THE NETHERLANDS.

D. Oostendorp... 322

Discussion 331

THE CURRENT PRACTICE OF COMMERCIAL CROSSBREEDING IN THE UK WITH PARTICULAR REFERENCE TO THE EFFECTS OF BREED CHOICE.

J.R. Southgate... 333

Discussion... 359

BEEF CROSSING IN THE DAIRY HERD - WHY DOES IT WORK IN THE BRITISH ISLES AND NOT IN GERMANY ?

H.J. Langholz... 360

Discussion... 372

CURRENT EFFICIENCY OF EMBRYO TRANSFER PROCEDURES AND THEIR POTENTIAL FOR INCREASING BEEF OUTPUT FROM THE DAIRY HERD.

J.M. Sreenan and M.G. Diskin 374

Discussion 384

GENERAL DISCUSSION 387

SUMMARY AND CONCLUDING REMARKS
E.P. Cunningham... 390

LIST OF PARTICIPANTS... 392

PREFACE

This publication contains the proceedings of a Seminar "Beef production from different dairy breeds and dairy beef crosses", held in Ireland on April 13-15, 1981, under the auspices of the Commission of the European Communities (CEC) as part of the European Communities (EC) programme for beef production research.

The CEC wishes to thank those representatives of Ireland who took responsibility for the organisation and conduct of this Seminar, notably Professor E.P. Cunningham, Dr. G.J. More O'Ferrall (local organiser), Dr. Patricia McGloughlin and Mr. R. Barlow. In particular, thanks are due to Dr. McGloughlin and Mr. Barlow for their recording of the discussions. Professor Ian Gordon of the Faculty of Agriculture, kindly made available the facilities of Lyons House, University College, for the Seminar.

Thanks are also accorded to the Chairmen of the Sessions : Professor D. Smidt, Dr. R.B. Thiessen, Professor A. Neimann-Sorensen, Professor E.P. Cunningham, and to all the participants who presented papers and took part in the discussions.

OBJECTIVES

The aims of the Seminar were to review recent comparisons of Holstein and Friesian strains with other dairy breeds for beef and veal production; to look at the use of beef breeds for crossing on dairy herds in various EEC countries, and to examine the economic and genetic balance between milk and beef traits in dual purpose bull testing and selection.

BACKGROUND

The rapid increase of Holstein genes in European Friesian populations in recent years, with a view to increasing milk production, has given rise to concern regarding the future potential beef production of the Friesian breed. The question of whether the use of beef crossing on a proportion of the dairy herds,as is practised in Great Britain, France and Ireland, could be extended to other countries as a means of increasing beef production could be a live proposition. As the number of cows is declining in many European countries, the question of whether greater economic weight should be put on beef traits in selecting replacement bulls needs examination, or should beef production in the future be concentrated on more specialised beef herds. Now that ova transplantation on a commercial scale is a live possibility, what are the prospects of increasing the beef production of dairy herds by these means ?

ADDRESS BY THE MINISTER OF STATE AT THE DEPARTMENT OF AGRICULTURE, MR. LORCAN ALLEN, OPENING THE EEC SEMINAR ON "BEEF PRODUCTION FROM THE DAIRY HERD"

Ladies and Gentlemen,

I would like at the outset to say how pleased I am to have been invited to open this seminar on Beef Production from the Dairy Herd. It is, I feel, particularly fitting that a seminar on this subject is being held in Ireland because beef from the dairy herd makes a massive contribution to our all important cattle industry. Beef cattle account for nearly 40% of our gross agricultural output and about 80% of this beef output is exported. Since about 75% of our beef cattle originates in the dairy herd the importance to us of the topic under discussion to-day is apparent.

The quality of beef from the dairy berd is largely influenced by two factors:-

- the beef merit of the dairy cow population, and
- the extent to which beef breed bulls are used for crossing on dairy cows and the merit of these bulls.

I am pleased to see that your programme over the next three days includes comprehensive reviews of both these important factors.

I know that in many parts of Europe there is a significant movement towards the breeding of more extreme dairy type cows, with poorer beef merit. However, a few years ago in Ireland, we carried out an extensive trial which compared the merits of these dairy types, including Canadian Holsteins and New Zealand Friesians, under Irish conditions. I know the results of this trial are being discussed here to-day so I won't go into them at any length. Briefly, while the Canadian Holstein crosses produced more milk, they also produced carcasses, 36% of which did not qualify for intervention.

I am aware that the current profitability of milk production can generate an interest in breeding cows of extreme dairy type, with little regard to their ability to produce calves suitable for the beef trade. While milk production must continually be improved, it should not be done at the expense of beef potential, or vice-versa. The aim must be to improve both. An added reason for taking account of the beef requirement is because female crosses from the dairy herd are the main source of cows for the suckler herd in Ireland.

Because of these considerations we are persisting with our long-standing national policy of breeding cows which are capable not only of efficient milk production, but also capable of producing calves which mature into good beef animals on our dominantly grass based feed. Indeed, I think that the current quality differentials in calf prices should prove a strong inducement to dairy farmers to breed better calves. In pursuing this policy we are, of course, fully conscious not alone of the interdependance of our own dairy and beef industries, but also the desirability of matching, as far as possible, Community production in both areas with market availability, so that costly surpluses whose existence is so often invoked as a criticism of the Common Agricultural Policy can be avoided.

Here in Ireland, in contrast to many other parts of Europe, a high proportion of our dairy cows are mated to beef breed bulls. This, I believe, makes good economic sense for us as it improves the overall quality of beef from the dairy herd and also leads to more selective breeding of dairy herd replacements. Choice of beef breed to cross on the dairy herd creates some problems for many farmers.

The faster growing, better conformation beef breeds tend to give rise to greater calving problems. Therein lies a challenge for your scientists to find a way to overcome the conflict between beefing merit and calving ease - by breeding perhaps, or by different management of cows, or some other techniques. I am sure you will have much to say on this topic and many other topics during your seminar.

In Ireland, only about 50% of our cows are bred by AI. This is a matter that concerns us. Since the best bulls are in AI the quickest way of improving the quality of our cattle is through the increased use of AI. The recent EEC special package of measures for Ireland, which was negotiated by the Minister for Agriculture, contains a subsidy for AI. It is hoped that this will be implemented shortly and the necessary increase in the use of AI will result.

Seminars such as yours are a necessary and valuable forum for exchange of ideas and provide an opportunity for identifying critical areas needing research and the coordination of research programmes. This can avoid unnecessary duplication of research effort and yield maximum returns from the limited resources which are available. I understand from my officials that the published proceedings of previous seminars in the series of which this is a part have made a very valuable contribution to dissemination and application of research findings.

I know there are many distinguished contributors to this seminar. Many have travelled long distances to share their knowledge. To them I extend a special welcome. You have a busy official programme over the next few days, but I trust you will find some time to relax and see and enjoy our country and its people.

Ladies and gentlemen, I have pleasure in declaring this seminar open.

TESTING OF DIFFERENT STRAINS OF FRIESIAN CATTLE IN POLAND. BEEF PERFORMANCE OF F_1 FRIESIAN BULLS FATTENED UNDER INTENSIVE FEEDING CONDITIONS TO 450 kg OR 550 kg OF LIVE WEIGHT

Z. Reklewski*

Institute of Genetics and Animal Breeding,
Polish Academy of Sciences, 05-551 Mrokow, Poland.

Summary

Beef progeny testing was conducted on animals obtained from mating local Black and White cows with Friesian bulls originating from the following countries : USA, Canada, Israel, New Zealand, Denmark, Sweden, Federal Republic of Germany, Great Britain, Holland and Poland.

Each group was evaluated genetically on the basis of the fattening results obtained from 28 bulls. A half of those animals was fattened up to 450 kg and the other half up to 550 kg.

During the fattening period the animals were fed individually on complete concentrates, rationed on the basis of the bull's body weight.

After slaughter the slaughter value was determined on the basis of a complete dissection of the right carcass side.

It was ascertained that the best fattening results (growth rate and feed conversion) were obtained by the progeny of the USA and Israel bulls. Significantly poorer results were obtained from the progeny of bulls from New Zealand and Denmark. The European types of Friesian cattle (Great Britain, Holland, Sweden and Poland) were characterized by good parameters of carcass quality: proportion of valuable cuts, dressing percentage and good musculature. The crossbreds after the USA and Israel bulls proved to be most valuable for intensive fattening.

Introduction

Different methods of utilising cattle and different methods of selection - applied in various countries - has meant that within the world population of Friesian cattle a series of local strains differing in production and exterior traits appeared. The greatest difference exists between the Holstein-Friesian strain from North America, selected for progress in milk traits, and the Dutch Black and White cattle utilised as a dual-purpose, dairy and beef breed. The American Holstein Friesian cattle are about 100 kg heavier and an average of 10 cm taller than Dutch cattle (Oldenbroek, 1977). The Dutch strain is characterised by a more compact build with shorter legs which gives the

* This paper was presented in Dr. Reklewski's absence

appearance of better musculature - these traits are considered important indicators of carcass quality and value.

In the last two decades the increasing interest in American Friesians has been observed in Europe. Numerous countries imported Holstein-Friesian cattle and semen to improve milk yield, growth capacity and udders of the local strains.

In this situation the objective comparison of the productive value and efficiency of particular strains became necessary and of international importance.

General outline

Nine countries - Canada, Denmark, the Federal Republic of Germany, Great Britain, Israel, the Netherlands, New Zealand, Sweden and the USA are taking part in the test.

FAO, which initiated this experiment, provides the international co-ordination and interchange of data.

Each country delivered semen in two batches of unproven bulls (17-20; 18-22 sires per country). In total, semen of 388 bulls was used in the experiment. The average number of doses per bull was 225 in the first batch and 250 in the second.

To obtain F_1 progeny only cows (contrary to primiparas) were inseminated in 70 state-owned farms. Inseminations of cows started in March 1974 and finished in December 1976.

Appreciating the role of Black and White cattle in the production of beef in European countries, also the beef value of the crosses was tested.

Material and Methods

The beef performance comparison of F_1 bulls under intensive feeding conditions is just a part of a larger scale test - the main part being held in the field.

The testing of the fattening and slaughter value was based on a sample of 28 bulls per each strain. Half of them were fattened to 450 kg live weight and the other half to 550 kg. At least fourteen sires were represented in each strain.

The F_1 calves were purchased at the age of about 5 weeks from the state farms involved in the crossbreeding project.

The actual testing period began when the animals were 110 days.

old.

During the fattening period there were three different stages in the nutrition plan :-

Stage 1 - Rearing of calves from 43 to 110 days - the animals received milk replacer, starter concentrate and hay.

Stage 2 - Between the 110th day and the body weight of 250 kg - concentrate No. 1 (12.18% of digestible protein and 1.21 Mcal of net energy for gain) and 1 kg hay per day.

Stage 3 - Between 250 and 450 or 550 kg body weight - concentrate No. 2 (11.0% digestible protein and 1.17 Mcal net energy for gain) and 1 kg hay/day.

For the rearing of calves the mean consumption of milk replacer amounted to 37 kg per animal.

The nutrient requirements of the fattened bulls were determined on the basis of the "Nutrient Requirements of Beef Cattle" National Academy of Sciences, Washington, 1970.

The diets were calculated individually for each animal, depending on its body weight. The body weight gains were controlled every fortnight. The leavings were weighed twice a week. On the basis of observations of the animals' appetite the diets were, if necessary, corrected every week.

The experimental animals were slaughtered on reaching 450 or 550 kg body weight. Prior to slaughter the animals were fasted for 24 hours. After slaughter the weights of the internal organs and of the internal, kidney and pelvis fats were recorded. The carcass quality was estimated on the basis of the contents of valuable cuts and main tissues in the right carcass side.

Results

In Table 1 the chosen preliminary results from the field comparison are presented. They are the least squares means of the weight at birth and at one year old and of the dairy gains during different stages of growth. The results cited are from the First Annual Report, May 27 1979 - May 26, 1980, Institute of Zootechnics, Cracow, Poland.

The least squares model included the effects of strain, sire within strain, year and season of birth and herd.

Table 1 Least squares means and standard errors of body weight and daily gains of F_1 bulls

Strain			No.	Live weight at[xx] Birth	Live weight at[xx] 12 Months	daily gains[x] 0 - 6	daily gains[x] 6 - 12
		LSM	5972	37.2	302	744	764
		Se		.28	2.2	6.0	9.7
A	USA	LSM	560	38.2	304	749	764
		Se		.39	3.1	8.4	13.5
B	Poland	LSM	1038	36.1	301	744	760
		Se		.39	3.2	8.4	14.0
C	Canada	LSM	580	37.9	304	748	766
		Se		.39	3.1	8.4	13.4
D	Denmark	LSM	562	37.6	301	735	765
		Se		.38	3.0	8.3	13.2
E	United Kingdom	LSM	498	36.8	298	737	754
		Se		.39	3.1	8.5	13.5
F	Sweden	LSM	561	37.4	307	750	780
		Se		.39	3.9	8.4	13.4
G	RFN	LSM	583	37.4	299	741	750
		Se		.38	3.0	8.3	13.3
H	Holland	LSM	519	36.5	295	730	739
		Se		.39	3.0	8.4	13.3
I	Israel	LSM	536	37.3	308	764	778
		Se		.41	3.3	8.8	14.3
K	New Zealand	LSM	535	36.8	304	738	786
		Se		.39	3.1	8.5	13.5

[x] Differences between breeds significant at $P \leqslant 0.05$

[xx] Differences between breeds significant at $p \leqslant 0.01$

The largest difference in birth weights between the F_1 USA and Polish bulls was 2.1 kg, i.e. 6% of the overall LS mean. The differences in live weight at the age of six months were smaller (4%) - the highest weights were recorded for the Israeli F_1 bulls.

In Table 2 are presented the results of fattening up to 450 kg. Already in the first stage of growth there occurred a clear differentiation between the genetic groups as regards body weight. In consequence, the mean body weight of animals at the beginning of the fattening period differed between groups. The heaviest bulls were those sired by Israeli (124 kg) bulls and USA sires (120 kg). Highly significantly lower body weights occurred in the following groups : New Zealand, Dutch, Swedish, English and significantly lower in the Danish and Canadian. The progeny of the USA sires grew best during the whole fattening period, reaching the average daily gain of 1207 g. Low daily gains were found in the New Zealand and Danish groups (1038 - 1059 g respectively).

High net gains were obtained by the crosses with the USA and Canadian sires - 635 g, while the lowest were in the New Zealand and Danish groups (576 g and 573 g).

The lowest intake of complete concentrate per unit of weight gain was observed in the progeny of the USA sires - 5.39 kg per 1 kg of live weight gain.

The worst feed conversion was found in the New Zealand group - 6.37 kg. Similar results were observed when comparing the feed conversion expressed in oat feed units.

As shown in Table 3 the highest dressing percentages were observed in the European strains (Dutch, Swedish, British) and in the New Zealand group. However, the differences between European Friesians and HF were not significant. No significant differences were found in the quantity of internal fat, content of valuable cuts and the MLD area, either.

As regards the results of the dissection (Table 4) the only clear differences were observed in the percentage of bone in the carcass side. A comparatively low share of bones was ascertained in the following groups : British - 17.04%. Polish - 17.06% and Dutch - 17.07%. The highest share of bones was

Table 2 Results of fattening F_1 bulls up to 450 kg of body weight

Strain	No.	Initial weight (kg)		Final weight (kg)		Days of fattening		Average daily gain (g)		Meat net gain (g)		Intake of concentrates per kg/gain (kg)		Intake of oat feed units per kg gain	
		$\bar{x}$	Sd	$\bar{x}$	Sd	$\bar{x}$	Sd	$\bar{x}$	Sd	$\bar{x}$	Sd	$\bar{x}$	Sd	$\bar{x}$	Sd
A	14	120.1	11.4	451.1	5.2	278.5	37.7	1207	158	636	75	5.39	0.71	5.59	0.71
B	14	115.7	11.8	450.4	7.4	294.1	39.0	1145	106	628	55	5.59	0.48	5.88	0.56
C	14	113.2	11.6	451.6	6.2	293.4	24.0	1156	95	636	37	5.77	0.52	5.98	0.53
D	14	113.3	13.4	448.4	5.9	322.1	36.6	1059	98	573	72	5.86	0.53	6.16	0.49
E	12	109.5	14.1	447.8	6.1	308.7	36.4	1106	99	613	41	5.73	0.66	5.98	0.65
F	14	107.5	9.6	448.1	7.1	302.7	24.4	1123	104	622	50	5.76	0.51	6.01	0.52
G	14	114.6	12.2	450.1	4.9	293.7	28.8	1154	96	620	50	5.62	0.57	5.85	0.57
H	13	109.8	9.9	450.3	8.3	305.9	29.4	1122	97	621	47	5.81	0.54	6.05	0.54
I	13	124.1	13.2	448.6	7.6	283.6	21.6	1151	76	633	40	5.69	0.35	5.93	0.33
K	15	105.0	9.2	445.5	5.6	330.6	32.8	1035	104	576	51	6.37	0.72	6.63	0.74
	A B-K^{x} I	E^{x} F^{x} H^{x} K^{xx} C^{x} D^{x} E^{xx} F^{xx} H^{xx} K^{xx}			K D	A^{xx} B^{xx} C^{xx} F^{x} G^{xx} I^{xx} A^{xx} B^{x} C^{x} G^{x} I^{xx}	K A E^{x} H^{x} D	A^{xx} B^{x} C^{xx} G^{xx} I^{x} A^{xx} C^{x} G^{x} I^{x}	K,D	A^{x} B^{x} C^{x} F^{x} G^{x} H^{x} I^{x}		A^{xx} B^{xx} C^{xx} D^{xx} E^{xx}	K K F^{x} G^{x} H^{x} I^{x}	A^{xx} B^{xx} C^{xx} E^{xx} F^{xx} G^{xx} I^{xx} H^{x} D^{xx}	D-A^{x}

Table 3 Results of post-slaughter analysis

Strain	No.	Weight of carcass side (kg)		Dressing percentage (%)		Weight of internal and kidney fats (kg)		Valuable cuts (%)		MLD area behind 13th vertebra (cm^2)	
		$\bar{x}$	Sd	$\bar{x}$	Sd	$\bar{x}$	Sd	$\bar{x}$	Sd	$\bar{x}$	Sd
A	14	124.1	3.8	55.99	1.87	16.9	3.3	63.29	1.91	68.6	7.7
B	14	126.2	4.3	57.01	1.31	17.7	4.6	62.67	1.24	71.1	9.1
C	14	125.5	4.4	56.98	2.02	18.8	2.9	62.53	0.85	66.8	10.2
D	14	123.3	3.3	56.41	1.62	17.9	4.0	62.64	0.94	68.3	6.4
E	12	125.9	4.4	57.29	1.63	17.7	4.6	62.81	1.06	69.6	9.4
F	14	125.3	5.4	57.37	1.76	17.1	5.5	63.05	0.66	72.8	11.8
G	14	124.1	3.9	56.31	1.07	15.6	1.9	63.09	0.64	71.5	10.7
H	13	126.3	3.2	57.45	1.16	18.7	3.3	62.90	1.34	73.8	8.6
I	13	124.4	3.7	56.55	2.27	17.0	3.3	62.75	1.04	71.0	8.0
K	15	124.4	3.0	57.31	1.17	18.6	4.3	62.25	1.36	65.8	9.1

Table 4 Results of the right carcass side dissection

Strain	No.	Lean in carcass side				Fat in carcass side				Bones in carcass side			
		kg		%		kg		%		kg		%	
		$\bar{x}$	Sd	$\bar{x}$	Sd	$\bar{x}$	Sd	$\bar{x}$	Sd	$\bar{x}$	Sd	$\bar{x}$	Sd
A	14	79.1	3.7	65.69	2.18	19.1	2.9	15.83	2.28	22.2	1.2	18.48	1.35
B	14	80.8	5.2	66.11	2.81	20.5	3.6	16.63	2.97	20.9	1.3	17.06	0.94
C	14	78.0	5.1	64.19	2.77	21.5	3.5	17.67	2.86	22.0	1.1	18.14	1.21
D	14	77.7	3.1	64.69	2.05	20.9	2.7	17.47	2.19	21.2	0.9	18.00	1.00
E	12	80.1	4.3	65.84	2.48	20.8	3.6	17.10	3.03	20.8	1.4	17.04	1.02
F	14	80.1	4.1	66.05	1.37	19.6	2.4	16.16	1.70	21.5	1.1	17.79	1.28
G	14	78.5	3.3	65.25	2.21	19.9	2.9	16.55	2.32	21.9	0.6	18.20	0.62
H	13	80.0	3.9	65.52	2.25	21.2	2.7	17.38	2.20	20.9	0.7	17.10	0.71
I	13	79.2	4.1	65.72	2.05	19.5	2.5	16.21	2.17	21.8	1.2	18.07	0.66
K	15	76.9	4.7	63.99	3.52	22.2	4.4	18.47	3.52	21.1	1.5	17.50	1.43

A < B^{xx}, D^{xx}, E^{xx}, H^{xx}, K^{x}

C,G < B^{x}, E^{x}, H^{x}

observed in the USA HF - 18.48%; the differences were highly significant.

On the basis of the results one may state that the progeny sired by the HF strains were characterized by the best growth and feed conversion and were most suitable for intensive fattening. However, the slaughter value and carcass quality were better in the groups of European strains.

To be able to rank the strains in a more precise way the estimation of the slaughter value was made based on two indicators :-

1) Average daily gain of lean meat $\dfrac{\text{carcass weight x \% of meat}}{\text{days of life}}$

2) Intake of oat feed units (FU) per 1 kg of lean meat $\dfrac{\text{total intake of FU}}{\text{weight of meat in carcass}}$

The results are shown in Table 5. In both cases the best strains were the USA HF, Polish and Israeli and the worst were Danish and New Zealand - the differences being highly significant.

The results obtained in the fattening trial up to 550 kg were similar to the results discussed previously. At the age of 110 days (Table 6) the greatest body weights reached the progeny of Swedish and Israeli sires. The highest daily gains were observed within the HF strains, i.e. from the USA, Israel and Canada. The difference between extreme groups reached 88 g, but was insignificant.

The lowest intake of concentrate per 1 kg of gain was observed in the progeny of Canadian sires - 6.11, Swedish and Danish groups had significantly worse feed conversion (6.71 and 6.70 respectively).

Table 7 presents the results of the post-slaughter analysis. The highest dressing percentage (59.57%) was obtained in the British group. Compared with this result the HF strains as well as New Zealand and Danish Friesians had poorer dressing percentage.

The weight of internal fat did not differ significantly among the strains. The share of valuable cuts was similar between groups with the exception of the New Zealand group.

As is seen in Table 8, the only significant differences in the carcass tissue content occurred in the case of bones -Polish

Table 5 Ranking of tested genotypes on the basis of an intensive fattening of F_1 bulls up to 450 kg of body weight

Strain	Daily gain of muscle tissue		Strain	Intake of oat units per kg lean gain	
	$\bar{x}$	Sd		$\bar{x}$	Sd
A	413	51	A	11.45	1.61
B	412	46	B	11.87	1.53
I	412	32	I	11.98	1.14
F	406	33	E	12.03	1.35
C	405	35	G	12.20	1.27
H	403	35	F	12.34	0.89
G	400	37	H	12.53	1.37
E	400	36	C	12.63	1.32
D	369	52	D	13.02	1.80
K	367	46	K	14.09	1.94

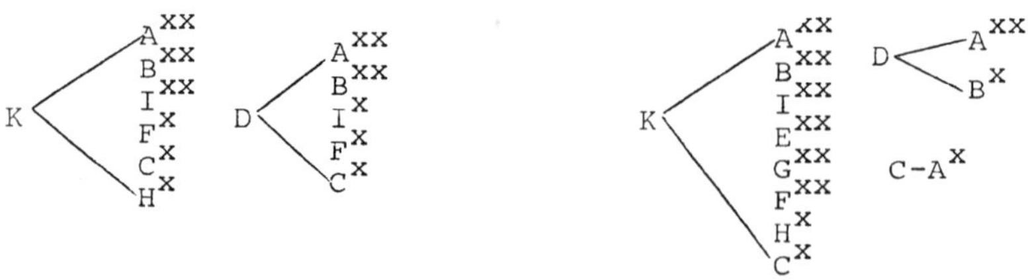

Table 6 Results of fattening F_1 bulls up to 550 kg

Strain	No.	Initial weight (kg)		Final weight (kg)		Days of fattening		Average daily gain (g)		Average net gain (g)		Intake of conc. per kg gain (kg)	
		$\bar{x}$	Sd	$\bar{x}$	Sd	$\bar{x}$	Sd	$\bar{x}$	Sd	$\bar{x}$	Sd	$\bar{x}$	Sd
A	13	111.8	14.9	543.7	4.8	362	38	1202	92	660	62	6.18	0.52
B	15	110.5	10.3	545.0	5.6	383	28	1140	80	641	37	6.57	0.49
C	14	106.9	11.0	543.8	6.1	375	33	1174	112	644	44	6.11	0.56
D	15	111.0	14.1	546.5	4.5	394	31	1113	84	633	42	6.70	0.55
E	16	110.1	11.0	544.9	6.5	391	32	1118	98	639	41	6.55	0.52
F	14	117.4	11.6	546.9	5.5	384	35	1120	103	648	48	6.71	0.52
G	14	107.6	12.2	546.4	7.5	387	30	1127	77	612	54	6.37	0.53
H	13	103.4	9.9	546.3	5.7	392	22	1131	53	640	37	6.51	0.30
I	14	117.9	16.2	540.5	7.6	368	24	1177	84	659	32	6.42	0.48
K	15	104.7	11.3	546.7	8.5	397	40	1122	106	618	59	6.40	0.50

F < K^{x}, H^{x}

I < C^{x}, H^{xx}, K^{x}

C < B^{x}, D^{xx}, E^{x}, F^{xx}

A < D^{x}, E^{x}, F^{x}

Table 7 Results of slaughter of F_1 bulls fattened up to 550 kg

Strain	No.	Weight of right carcass side (kg)		Dressing percentage (%)		Internal & kidney fat (kg)		Valuable cuts (%)		MLD area	
		$\bar{x}$	Sd	$\bar{x}$	Sd	$\bar{x}$	Sd	$\bar{x}$	Sd	$\bar{x}$	Sd
A	13	153	5	57.80	1.50	26.7	6.3	62.06	1.10	76.5	8.4
B	15	157	5	58.56	1.57	27.9	6.0	61.80	1.05	78.1	8.8
C	14	154	4	57.74	1.68	27.3	5.5	62.54	0.59	76.3	8.6
D	15	156	4	58.18	1.50	26.3	8.5	62.21	1.02	79.7	11.5
E	16	159	4	59.57	1.22	28.2	6.7	62.36	0.90	82.1	6.5
F	14	158	3	58.98	1.13	26.8	5.2	62.13	P½97	75.6	8.5
G	14	157	4	58.50	1.60	26.4	4.4	62.56	1.20	80.8	9.3
H	13	159	4	59.11	1.92	24.7	4.2	62.17	0.77	82.5	9.9
I	14	157	4	58.34	1.65	28.7	5.0	61.99	1.06	82.1	9.2
K	15	155	4	57.82	1.22	27.2	5.7	60.99	0.93	81.6	9.1

A: B^{xx} E^{xx} F^{xx} G^{x} H^{xx} I^{x}
K: E^{x} H^{x}
C: E^{x} F^{x} H^{xx}

E: A^{xx} C^{xx} B^{x} I^{x} K^{xx}
H: A^{x} C^{x} K^{x}

K: A^{xx} B^{x} C^{xx} D^{xx} E^{xx} F^{xx} G^{xx} H^{xx} I^{xx}

Table 8 Dissection results obtained from carcass sides of F_1 bulls fattened up to 550 kg

Strain	No.	Lean in carcass side				Fat in carcass side				Bones in carcass side			
		kg		%		kg		%		kg		%	
		$\bar{x}$	Sd	$\bar{x}$	Sd	$\bar{x}$	Sd	$\bar{x}$	Sd	$\bar{x}$	Sd	$\bar{x}$	Sd
A	13	94.9	4.3	64.47	1.36	27.0	2.2	18.36	2.31	25.1	1.3	17.17	1.68
B	15	96.3	6.0	63.95	2.73	30.1	4.4	20.03	2.90	24.1	1.1	16.02	0.83
C	14	93.9	4.5	63.60	2.00	27.8	2.8	18.84	1.84	25.9	1.5	17.56	1.05
D	15	94.8	4.7	63.42	2.51	29.6	3.9	19.82	2.68	25.1	1.2	16.76	0.73
E	16	97.8	6.0	64.05	3.31	30.0	5.1	19.67	3.41	24.8	1.5	16.28	0.96
F	14	97.8	4.8	64.57	2.54	28.6	3.8	18.83	3.24	25.1	1.1	16.60	1.14
G	14	97.0	5.3	64.00	2.29	28.0	3.3	18.79	1.34	25.8	1.5	17.21	2.55
H	13	98.5	6.5	63.97	3.12	29.6	4.0	19.13	2.73	25.2	1.3	16.90	2.98
I	14	95.8	4.6	63.72	1.79	29.5	3.0	19.69	2.05	22.2	1.6	16.59	1.34
K	15	93.1	6.0	62.55	3.28	31.2	4.6	20.97	3.12	24.5	1.5	16.43	1.09

B < C^{xx}, G^{xx}, H^{x}

K < C^{x}, G^{x}

and New Zealand strains had the lowest content of bones.

As regards growth rate the best proved to be F_1 bulls sired by Holstein-Friesian sires. The qualitative evaluation of the carcasses demonstrated that, during intensive fattening to higher body weights, European strains of Black and White cattle reached a greater degree of fatness than did the crosses with HF sires.

Ranking the strains on the basis of "lean meat" gain (Table 9) the highest places fall to the progeny of the USA sires - 423 g, Israeli sires - 416 g and Swedish sires - 413 g. The progeny of German and New Zealand sires showed a slow rate of meat gain (395 and 384 g respectively). Accepting the intake of nutrients per 1 kg of "lean meat" gain as criterion of classification, the first three places fall to the groups of progeny of the USA (14.06 kg), Canadian (14.14 kg) and Israeli (14.28 kg) sires. New Zealand and Danish strains proved again to be the least effective, although, in this case, the differences were not significant.

Both fattening tests were conducted according to the same methods. The only difference was in the length of the fattening period. On the other hand, the results showed that there was no interaction between the strains and fattening trials. In order to obtain possibly complete results of the slaughter value of crosses and to check on the possible interaction, it was decided to combine the results from both replicas. The material was analysed according to a least squares model involving the effects of strain, trial and strain x trial interaction; the calculations covered chosen performance parameters, selected so as to eliminate as far as possible the influence of the different slaughter weights of the animals. Again, the best growing strains (Table 10) proved to be the Holstein -Friesians, while the Danish and New Zealand strains remained inferior. The European breeds had, in general, significantly better carcass composition (Table 11). Only the HF from the USA reached the top in lean content.

Table 12 presents the ranking of strains on the basis of the accepted indicators of slaughter value.

When using the mean daily gain of "lean meat" as the criterion

Table 9 Ranking of genotypes on the basis of an intensive fattening of F_1 bulls up to 550 kg

Strain	Average daily gain of muscle tissue (g)		Strain	Intake of oat feed units per kg lean gain (g)	
	$\bar{x}$	Sd		$\bar{x}$	Sd
A	423	11	A	14.06	0.39
I	416	11	C	14.14	0.38
F	413	10	I	14.28	0.38
H	408	11	G	14.30	0.38
E	406	10	F	14.62	0.38
B	406	10	H	14.62	0.39
C	405	10	B	14.67	0.37
D	397	10	E	14.76	0.36
G	395	10	D	15.28	0.37
K	383	10	K	15.37	0.36

Table 10 Ranking of genotypes on the basis of results of fattening. Least squares means from overall analysis of variance

Mean daily gain (g)			Mean net gain (g)			Intake of complete concentrate per 1 kg of gain		
Strain	$\bar{x}$	Se	Strain	$\bar{x}$	Se	Strain	$\bar{x}$	Se
A	1204	19	A	648	10	A	5.78	0.10
C	1167	19	I	646	10	C	5.94	0.10
I	1164	19	C	640	9	G	5.99	0.10
B	1142	18	B	636	9	I	6.05	0.10
G	1141	19	F	635	9	B	6.08	0.10
H	1126	19	H	630	10	H	6.16	0.10
F	1125	19	H	626	9	E	6.19	0.10
E	1112	19	G	616	9	F	6.24	0.10
D	1086	18	D	603	9	D	6.28	0.10
K	1080	18	K	597	9	K	6.42	0.10

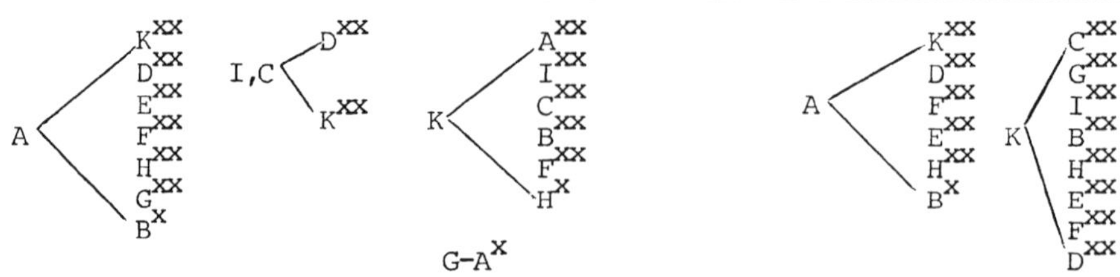

Table 11 Ranking of strains on the basis of L.S. means of dissection results

Dressing percentage			Valuable cuts %			Lean in carcass side %			Fat in carcass side %			Bones in carcass side %		
Strain	$\bar{x}$	Se	Strain	$\bar{x}$	Se	Strain	$\bar{x}$	Se	Strain	$\bar{x}$	Se	Strain	$\bar{x}$	Se
E	58.43	0.28	G	62.82	0.19	F	65.30	0.48	A	17.10	0.47	B	16.53	0.25
H	58.27	0.29	F	62.59	0.19	A	65.08	0.48	F	17.49	0.48	E	16.67	0.25
F	58.18	0.28	E	62.58	0.19	B	65.05	0.47	G	17.63	0.48	H	17.01	0.26
B	57.79	0.28	H	62.57	0.20	E	64.95	0.48	I	17.96	0.47	K	17.01	0.25
K	57.60	0.27	C	62.53	0.19	H	64.74	0.49	C	18.25	0.47	F	17.21	0.25
I	57.44	0.29	I	62.48	0.19	I	64.72	0.48	H	18.26	0.47	I	17.32	0.25
G	57.41	0.28	D	62.43	0.18	G	64.58	0.48	E	18.38	0.48	D	17.39	0.25
C	57.36	0.28	A	62.42	0.19	D	64.02	0.47	B	18.42	0.49	G	17.79	0.25
D	57.27	0.28	B	62.24	0.18	C	63.90	0.48	D	18.59	0.46	A	17.82	0.26
A	56.59	0.29	K	61.62	0.18	K	63.27	0.46	K	19.72	0.45	C	17.85	0.25

Dressing percentage: E – A^{xx}, D^{xx}, C^{xx}, G^{x}, I^{x}; H – A^{xx}, D^{xx}; F–A^{xx}

Valuable cuts: K – G^{xx}, F^{xx}, E^{xx}, H^{xx}, C^{xx}, I^{xx}, D^{xx}, A^{x}

Fat in carcass side: K – A^{xx}, F^{xx}, G^{xx}, I^{x}; D–A^{x}

Bones in carcass side: B – C^{xx}, A^{xx}, G^{xx}, D^{x}; G – K^{xx}, H^{xx}; E – C^{xx}, A^{xx}, G^{xx}

Table 12 Ordering of strains on the basis of a beef utility test of F_1 bulls fattened intensively. Mean of least squares analysis (independent of the final body weight of bulls - 450 or 550 kg)

Strain	No.	Mean daily gain of "lean meat" (g)		Strain	No.	Intake of oat feed units per kg of "lean meat" gain	
		$\bar{x}$	Sd			$\bar{x}$	Sd
A	27	417.9	7.6	A	27	12.76	0.27
I	27	414.0	7.6	I	27	13.13	0.27
F	28	409.5	7.5	G	28	13.25	0.27
B	29	409.2	7.3	B	29	13.27	0.26
C	28	405.4	7.5	E	28	13.39	0.27
H	26	405.3	7.7	C	28	13.39	0.27
E	28	403.2	7.5	F	28	13.48	0.27
G	28	397.6	7.5	H	26	13.57	0.28
D	29	383.3	7.3	D	29	14.14	0.26
K	27	375.5	7.2	K	30	14.73	0.26
	230	402.1	2.4		280	13.51	0.08

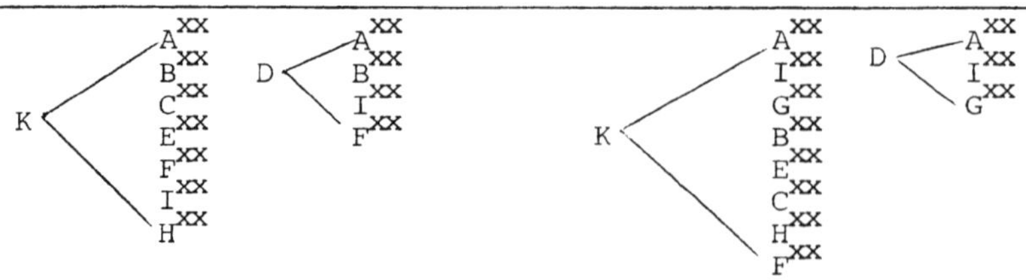

of the slaughter value it was observed that the New Zealand and Dutch strains received a significantly lower marking when compared with the USA, Israeli, Swedish, Polish, English and Dutch strains. The ranking of the strains was not subjected to greater changes when the classification was based on the intake of nutrients per 1 kg of "lean meat" gain.

It is also worthwhile to state that only in one case was the interaction between strain and trial found to be significant. It happened in the case of the intake of complete concentrate per 1 kg of weight gain.

It was unexpected that the best results in a slaughter value test would be obtained by bulls sired by HF sires from USA and Israel. Those populations had been selected exclusively for milk production traits. This may confirm Lush's (1961) opinion that selection directed at the improvement of one trait may, in reality, be a selection for an index covering many traits due to correlated responses. Of course, one cannot, evaluating the results obtained from generation F_1, exclude the influence of heterosis, which would make it difficult to estimate the beef value of the tested strains. The final determination of their genetical value will only be possible when the results obtained for generation R_1 are available.

References

Lush, J.L. 1961. Selection indexes for dairy cattle. Zeitschrift fur Tierzuchtung und Zuchtungsbiologie 75, 249-261.

Oldenbroek, J.K. 1977. Vergleich nordamerykanischer Holstein Friesians mit niederlandischer Schwarz und Rotbunten. Tierzuchter 9, 29, 374-378.

COMMENTS ON DR. REKLEWSKI'S PAPER

Chairman: Prof. Dr. D. Smidt

Thiessen: Results from breed comparisons have shown that the greatest variation in production traits, particularly food efficiency, is observed when comparisons are made at the same bodyweight. In view of the influence of differences in mature size, I would like to see comparisons made at the same age, and also by regression analysis to the same degree of fatness.

Langholz: Results from the Polish trial have been obtained on heavy concentrate diets, and hence it is not surprising that they largely reflect differences in the mature size of the Friesian strains. Reducing the nutrient concentration of the diet, or limiting the time on test, might be expected to change the ranking.

Foulley: Did the within-strain variation vary between the different strains ?

More O'Ferrall: The number of animals per sire was too small to allow a satisfactory answer to this question.

Cunningham: The data come in two parts: a large trial (6,000 head) giving growth data, and a limited trial (15 per group) giving carcass data. It is notable that the large trial showed no differences of economic consequence in growth performance between strains. One surprising result in the carcass trial was that the New Zealand strain produced the fattest carcasses.

COMPARISON OF BEEF PERFORMANCE OF DIFFERENT FRIESIAN STRAINS FROM PROGENY TEST RESULTS IN IRELAND

J. Flanagan

Department of Agriculture, Kildare Street, Dublin 2, Ireland

Summary

The Friesian breed in Ireland is mainly of British origin but with importation of strains from other countries from time to time. An analysis of beef progeny test results in Ireland gave an evaluation of the relative merits of different Friesian strains for beef production. German Friesians were similar to Irish Friesians. New Zealand Friesians were slower growing with inferior carcass conformation but with similar carcass composition to Irish Friesians. Canadian Holstein-Friesian progeny were equal in growth rate to Irish Friesian but were substantially inferior in carcass conformation. The Canadian carcasses had significantly more bone and less fat than the Irish strain, giving about the same percentage of saleable meat for the two strains.

Introduction

The Friesian breed in Ireland is a dual purpose breed. Almost all cows used for milk production are now Friesian, and the majority of the calves for beef production come from the Friesian herd as pure Friesians or as Friesian crosses. The selection objective for the Friesian breed is to improve simultaneously both its dairy and beefing qualities.

The origin of the Irish Friesian is mainly from Great Britain, with small importations of bulls or semen from other countries from time to time. The more recent non-British importations include Dutch Friesian in 1964, German Friesian in 1970, Canadian Holstein and New Zealand Friesian in 1973/75 and Dutch Friesian in 1980.

Friesian bulls in AI in Ireland are generally progeny tested for beef traits as well as for dairy traits. Beef progeny test results of bulls of different strains give an evaluation of the relative merits of the strains for beef production. This paper is concerned with such an analysis of beef progeny test results in Ireland.

Materials and Methods

Beef progeny testing in Ireland is carried out by the AI stations under the general supervision of the Department of Agriculture. A random sample of about 20 male progeny per bull out of Irish Friesian cows is purchased and reared together in groups on AI station or co-operating farms under commercial beef production conditions. Calves are reared on milk substitutes and meals and thereafter fed a mainly grass based diet to produce finished steers for slaughter at about 2 to 2½ years of age. Animals from each farm are slaughtered in one or two batches depending on finish, with equal proportions of progeny of each bull in each batch. At slaughter, carcasses are classified according to the national beef carcass classification scheme, for conformation and fatness. There are seven conformation classes, denoted I, R, E, L, A, N and D, which were converted to the numberical scale 7, 6, 5, 4, 3, 2 and 1, for analysis purposes. Best conformation is denoted by 7 and poorest conformation by 1. There are also 7 fat classes denoted 1, 2, 3, 4, 5, 6 and 7, where 1 is least fat cover and 7 is most fat cover. (See reference No. 1).

At slaughter carcasses are also graded for intervention into four acceptable grades denoted 1, 2, 3 and 4, and one reject grade. Intervention buying-in price is highest for grade 1 and least for grade 4.

In 1978, as part of a Friesian strain comparison trial, a sample of carcasses of Irish, Canadian and New Zealand strain progeny were cut up to estimate carcass composition. The right hand side of each carcass was deboned and cut into standardised fat trimmed cuts according to a specification for the continental vac-pac trade (Ryan, 1978).

The data analysed for this paper consisted of 5 measurements - carcass weight, length, killout %, conformation and fatness on the progeny of bulls of Irish, Canadian and New Zealand strains and the cut up data on progeny of Irish, Canadian and New Zealand strains from the 1978 trial.

Data were analysed by the least squares method using a model with effects for farm-batch, age at start of test, strain and bull within strain. Results are generally expressed in the form of least squares means and their standard errors.

Results

The performance of progeny of Canadian, New Zealand and German Friesian strains relative to Irish Friesian is given in Table 1 for five carcass traits - weight, length, killout percentage conformation and external fatness. Canadian strain progeny were about the same weight and had similar killout percentage to Irish Friesian. The Canadian strain carcasses were significantly longer with poorer conformation and less fat.

New Zealand strain progeny were significantly lighter than their Irish contemporaries but were about the same length and gave similar killout percentages. New Zealand conformation was significantly poorer than Irish, but external fat cover was similar for the two strains.

German strain progeny were not significantly different from the Irish for any of the five traits.

The percentage of progeny of the Irish, Canadian and New Zealand strains falling into the different intervention grades is given in Table 2. Of the Canadian progeny, 36% were rejects for intervention compared with 19% of Irish Friesian and 25% of the New Zealand progeny. Of the carcasses that qualified for intervention, the percentages in the top grade were 44, 18 and 27 for the Irish, Canadian and New Zealand strains respectively.

Carcass composition in terms of percentage saleable meat, percentage fat trim and percentage bone is given in Table 3 for the Irish, Canadian and New Zealand strains. Canadian progeny had significantly more bone and significantly less fat trim than the Irish progeny, giving about the same percentage saleable meat for the two strains. Carcass composition of the New Zealand strain was similar to that of the Irish strain.

Table 1 Average beef progeny test results for Irish, Canadian, New Zealand and German Friesian strains

	Irish	Canadian	New Zealand	German
No. of bulls	86	10	4	10
No. of progeny	1624	244	124	141
Carcass weight kg	284.9 ± 0.6	282.0 ± 1.9	275.9* ± 2.0	284.9 ± 1.9
Carcass length cm	130.9 ± 0.1	132.3* ± 0.3	130.4 ± 0.3	130.4 ± 0.3
Killout %	54.13 ± 0.04	53.64 ± 0.14	53.85 ± 0.15	54.40 ± 0.15
Carcass conformation score	4.20 ± 0.02	3.86** ± 0.06	3.96** ± 0.06	4.28 ± 0.06
Carcass fatness score	3.00 ± 0.02	2.71* ± 0.06	2.98 ± 0.06	-

* Significantly different from Irish Friesian at $P < .05$ level

** Significantly different from Irish Friesian at $P < .01$ level

Table 2 Intervention grading of Irish, Canadian and New Zealand strains of Friesians

% of strain in Grade

	Irish	Canadian	New Zealand
Grade 1	36%	11%	20%
" 2	35	42	44
" 3	9	10	11
" 4	1	1	0
Reject	19	36	25

Discussion

The relative merits of the Canadian Holstein and European Friesian strains found in this study are in general agreement with results of comparison of these strains elsewhere. The limited number of published trials prior to 1974 were summarised by Cunningham (1974). A comparison of pure British Friesian

Table 3 Carcass composition of Irish, Canadian and New Zealand strains of Friesian

	Irish	Canadian	New Zealand
Number of bulls	8	2	4
Number of carcasses	126	64	72
% saleable meat	69.29 ± .13	69.19 ± .19	69.25 ± .18
% trim	11.66 ± .17	10.99* ± .24	11.77 ± .22
% bone	19.02 ± .09	19.80** ± .14	18.96 ± .13

* Significantly different from Irish and New Zealand Friesians at $P < .05$ level.

** Significantly different from Irish and New Zealand Friesians at $P < .01$ level.

with pure Canadian Holstein steers for beef production from an 18-month cereal/grass system reported by Cook and Newton (1979) showed a small but significant difference in carcass gain in favour of British Friesians. The British Friesian was clearly superior in carcass conformation and had 0.7% more saleable meat and 1% less bone. Another comparison showed similar growth rates on performance test for British Friesian and Canadian Holstein crosses on Norwegian NRF cattle (Skjervold and Odergard, 1979).

The results of this study are not in good agreement with the preliminary report on the relative beef merit of these strains from the International Friesian strain comparison in Poland. In the Polish trial the New Zealand Friesian seems to have growth rates similar to the North American strains and about 5% superior to the German and British strains. This is in contrast to the present study where the Canadian growth rate was similar to the Irish (British) and German strains but these were superior to the New Zealand by about 30%.

References

Beef Carcase Classification Scheme. Leaflet of the Department of Agriculture, Dublin.

Cook, K.N. and Jennifer M. Newton, 1979. A comparison of Canadian Holstein and British Friesian steers for the production of beef from an 18-month grass/cereal system.

Cunningham, E.P. 1974. Comparisons of North American Holstein and European Friesian cattle for dairy and beef production. An Foras Taluntais, Dublin (mimeograph).

Ryan, P.O. 1978. Classification and carcase value. CBF - Irish Livestock and Meat Board 6th International Beef Symposium, Dublin 1978.

Skjervold, H. and Odergard, A.I. 1979. Preliminary results from the use of Canadian, British and USA Friesian bulls in the Norwegian Red cattle population. Paper read at the 38th Annual Meeting of the EAAP, Harrogate, England.

DISCUSSION ON DR. J. FLANAGAN'S PAPER

Chairman: Prof. Dr. D. Smidt

Southgate: Could you expand on the slaughter criteria for progeny-test groups; was it based on age, weight, fatness or a combination of these ?

Flanagan: Animals were reared commercially and purchased when fit for slaughter at approximately two years of age, based on a commercial decision. If animals were not fit for slaughter, then a proportion from each bull was slaughtered.

Thiessen: Could you say something more about the importance of conformation. There was considerable variation in this trait, but not in percentage muscle.

Flanagan: The Irish trade places a lot of importance on conformation. No differences were found here, although previous dissections did detect small differences in composition relative to conformation.

Thiessen: Will the influence of conformation on price change in the future ?

Flanagan: There is no evidence that it has changed in the past.

More O'Ferrall: Would you care to expand on the economic return to the commercial producer of Holstein crosses, compared with Friesian crosses in situations where the factory is vacuum packing the beef, compared to selling carcasses.

Flanagan: Factories do not pay producers any more for Holstein crosses if they are vacuum packing. So you could say that if they were to purchase Holstein crosses for a vacuum packing trade at a discounted price, they could make more profit because of relatively small differences in percent of saleable meat.

Langholz: Is there any information available about the dams to which bulls were mated. How was it assured that no bias was possible by selective matings between cows within farms, and/or between regions ?

Flanagan: No bias was possible. The cows to which proven and young unproven bulls are mated is predetermined, and then selection within groups is random. All animals were purchased subsequently, unless the price was extremely high.

Langholz: Was there any possibility of selective feeding of animals by farmers ?

Flanagan : Standards are set and monetary premiums are paid to ensure that these standards are kept, so the chances of selective feeding are remote.

All dams were Irish Friesian, if this was not clear in the paper.

Fewson: Were the different breeds introduced in different years, and did this introduce biases due to timelag effects, etc. The latter may be expected if there was upgrading in the source populations.

Flanagan: Yes, German sires were compared in 1971 and Canadian sires compared in 1975; so this could lead to small biases.

There is an error in my paper in relation to the performance of New Zealand bulls in the Polish trial. Paper I in this Conference indicated that the performance of New Zealand sires was relatively poorer than I had been led to believe from preliminary reports from that trial.

DANISH BREEDING STRATEGIES FOR MILK AND BEEF TRAITS IN DAIRY AND DUAL PURPOSE CATTLE BREEDS

B. Bech Andersen

National Institute of Animal Science,
Rolighedsvej 25, 1958 Copenhagen V, Denmark

Summary

The breeding goal for Red Danish and Danish Friesian is a simultaneous improvement of milk yield, growth rate, feed efficiency and carcass quality.

The breeding work is strongly influenced by the import of Holstein Friesian genes into the Danish Friesian and American Brown Swiss genes into the Red Danish. A great variation in beef traits has been demonstrated among the imported genes, and, therefore, a systematic testing and selection of the crossbred potential breeding bulls are important.

In Denmark such selection is mainly based on performance tests of growth rate, appetite and ultrasonic muscle area (600 bulls/year), and to a lesser extent on progeny tests for growth rate, carcass composition and meat quality (30 progeny groups/ year).

Introduction

The Danish production of milk and beef is mainly based on dual-purpose breeds selected for both traits. On average 2/3 of the income of cattle farmers is based on milk production and 1/3 on beef production. From a biological point of view a simultaneous milk/beef production on dairy or dual-purpose cattle is the most efficient production form, as the maintenance requirements of a dairy cow at the same time is the basis for milk production, foetus production and the cow's own growth until an age of four years. It indicates an increasing importance of combined milk/beef production, not necessarily only on the same farms, but with a beef production based on culled cows and bull calves from cows primarily kept for milk production.

The combined milk/beef production can be based on beef x dairy crossbreeding or dual-purpose breeds. As shown by Liboriussen (1981), beef x dairy crossbreeding is not expected to be of great importance in Denmark in the future. It will actualize the dual-purpose type of cattle.

By including growth rate and feed efficiency in the breeding scheme it will reduce the costs of food, labour and investment in the production of fattening bulls and replacement heifers, and possibly it will also increase the efficiency of the dairy cows. By including carcass traits in the breeding scheme it will increase the value of young bulls, surplus heifers and culled cows.

Biological factors and breeding structure

Danish cattle breeding is favoured by relatively good technical facilities. About 90% of the cows are served by AI and 70.7 % are milk recorded. It means that the active breeding unit is 64% of the cows or 650,000 cows in total. In addition 600 potential bulls for AI can be performance tested on central stations each year, and within a few years a new station with 300 places will be added to the capacity.

In total 7 mia. Scand. f.u. per year are used in the Danish cattle industry. Fifteen per cent are used for young bull production, 25% for replacement heifers, 25% to meet the maintenance requirement of the cows and 35% directly for the milk production. Feed cost is the most important part of the variable costs in the production, and, therefore, a genetic improvement in feed conversion efficiency can significantly improve the economic results in both milk and beef production. Also the genetic variation in appetite of the animals is important, as most of the high yielding cows are not able to eat enough to meet the energy and protein requirements in the first part of the lactation.

Growth rate in young animals (e.g. young bulls on performance test) is largely a function of appetite (physically and/or physiologically regulated); the lean tissue growth capacity and the residual efficiency (maintenance requirement and efficiency of the digestion tract) (Andersen et al. 1981a). In preliminary Danish investigations we have seen that the genetic parameters depend on the testing system. In tests with restrictive feeding we found a genetic variation in growth rate on 45 g/day. In an ad lib. system, when appetite is included in the test results, the genetic variation is increased to 70 g/day approx. And the genetic correlation between growth rate and feed/gain is -0.95 in the restrictive system and -0.65 in the ad lib. system. As a consequence we have changed the feeding regime from a restrictive system to an ad lib. feeding, with a roughage mixture based on treated straw and molasses. Individual recording of feed intake is included in the system, and as a result of the growth rate selection we expect to improve the physical appetite, feed efficiency and maintenance requirement in our breeds.

It is shown that both genetic improvement of milk production capacity and import of genes from dairy populations in North America lead to a decrease in carcass value of our cattle (Andersen & Andersen,1975, and Andersen, 1981b). In addition a one-sided selection for daily gain causes an indirect reduction in dressing percentage and lean/bone ratio. However, by measurements of ultrasonic loin eye area on the performance tested bulls, and by combining that area with daily gain into a selection index, it is possible to improve growth capacity and feed efficiency without an adverse effect on carcass quality (Table 1). Such a selection also causes an increase in birth weight and mature cow weight. It is important to notice, however, that these increases do not affect the ratio between birth weight and mature cow weight.

Breeding strategies

In 1974 an official breeding plan for the Danish cattle breeds was published (Petersen et al. 1974), and that plan is followed by most of the National Breed Associations and local AI Centres. However, import of genes (especially semen) from other cattle populations has been very widespread and important during the last years.

(a) Import of genes

The breeding work with Black and White Danish cattle has for years been based on import of bulls from other European countries. During the last 10 years, however, import of semen from Holstein-Friesian in North America has been very popular, and in 1980

Table 1 Expected genetic long term effect of performance test selection based on indexes of different construction (Andersen, 1978)

	Genetic response in:														
	Daily gain in g of :					Scand. f.u. per									
Index	Live wt.	Carcass wt.	Lean	Fat	Bone	Kg gain	Kg lean	LD	D%	L/B	L/F	LP/LT	GL	BW	Mat. cow wt.
DG	100	36	36	-12	10	-0.4	-1.2	-1	-1.5	-0.15	0.7	0.4	1.2	3.3	48
LD	-5	23	30	-2	-4	-0.0	-0.8	11	1.7	0.35	0.4	0.6	-0.3	0.9	-8
DG+LD	52	36	37	-5	3	-0.3	-1.3	9	0.5	0.20	0.7	0.7	0.5	2.7	19
DG+(LD) [1]	90	41	37	-6	9	-0.5	-1.3	4	-0.8	0	0.8	0.6	1.0	3.5	40
DG+GL	90	37	31	-5	10	-0.5	-1.2	-1	-1.0	-0.09	0.7	0.5	5.6	5.3	82
DG+(GL) [2]	75	22	17	-5	9	-0.3	-0.6	0	-1.5	-0.17	0.5	0.1	-4.1	0	-7
DG+LD+(GL) [2]	9	20	24	-3	-2	-0.1	-0.7	10	1.1	0.28	0.4	0.6	-2.7	0	-25
DG+LD+(GL) [3]	77	17	11	-4	10	-0.3	-0.4	-3	-2.0	-0.26	0.4	0	-3.5	0	0

[1] restricted index with lean/bone ratio constant.
[2] restricted index with birth weight constant.
[3] restricted index with birth weight and mature cow weight constant.

Scand.f.u. = Scandinavian feed units
LD = Longissimus dorsi muscle area
D% = Dressing percentage
L/B = Lean/Bone ratio
L/F = Lean/Fat ratio
LP/LT = 100 x lean in pistol/total lean
GL = Gestation length
BW = Birth weight

most of the AI bulls were graded up with Holstein-Friesian (Table 2). The proportion of Holstein-Friesian genes in the semen used in 1979 was on average 34% and the proportion of Holstein-Friesian genes in the Black and White cows was, on average, 9%. This proportion will increase enormously in the years to come. The importation has increased the variation in the Danish Black and White cattle population and results from our performance tests presented in Table 3 show that some of the North American bulls are excellent in growth rate and muscularity, whereas other bulls show very poor results for these traits. It illustrates the importance of a systematic testing and selection among the imported genes.

Table 2 The use of Holstein-Friesian genes in the Danish Black and White cattle population (Madsen, 1981)

		% H-F genes :			
Year	% Young bulls with H-F genes	Young bulls	Semen used	Calves born	Dairy cows
1970	5	2	3	1	0
1971	3	1	2	1	0
1972	5	2	2	1	0.3
1973	10	5	5	1	0.6
1974	8	4	3	3	0.8
1975	22	15	8	2	0.9
1976	50	23	14	5	1.6
1977	71	34	19	8	1.8
1978	86	45	25	11	2.8
1979	93	56	34	15	4.4
1980	?	?	?	20	6.6
1981	?	?	?	?	9.2

The Red Danish has been a closed purebred breed for more than 100 years. The butterfat and protein production capacity of the breed is equal or even higher than the Friesians, but a high degree of inbreeding has led to relatively poor fertility, high

Table 3 Performance test results of siregroups with different proportions of H-F genes (Andersen et al. 1981b)

Name of sire	% H-F genes	No. of sons	Average of siregroups in					
			Index				Weight in kg at	
			Gain	Muscle area	Daily gain	Muscle area	1½ mts	11 mts
SDJ Tema	50	16	104	101	1329	61	64	455
VAR Arli	50	22	101	102	1244	64	72	437
MJY Black	50	77	99	103	1221	63	64	423
ARL Chief	100	28	101	101	1275	63	66	441
Winfarm	100	38	102	100	1291	60	66	445
MJY Frans	0	30	100	102	1248	63	66	433
VIB Ernst	0	10	101	101	1258	60	64	433
GJO Oskar	50	56	100	101	1234	63	65	427
ALB Frans	0	10	101	100	1263	59	69	440
P.C. Star	100	12	104	98	1321	56	68	456
Starchief	100	54	102	99	1274	59	69	443
Stern	100	92	100	100	1233	59	66	428
Gay Ideal	100	17	99	99	1234	60	66	429
QUA Ultimate	100	36	102	96	1289	55	68	447
AST Galaxy	100	19	98	98	1198	59	60	412
M.R. Design	100	20	98	98	1190	57	63	413

calf mortality etc. In 1979 it was, therefore, decided to start a grading up scheme with import of semen from American Brown Swiss. The plan was based on experimental results from controlled import of semen from the Finnish Ayrshire, Swedish Red and White, the Dutch breed MRIj, Red Holstein and American Brown Swiss breeds. Computer simulation studies showed that during a five-year period 50% of the bull dams and 5-10% of the remainder cows should be inseminated with imported Brown Swiss semen. It was found that after 20 years more than 90% of the Red Danish cows should contain more or less Brown Swiss genes, and on average 20-25% of all genes in the population should be Brown Swiss. It is required that all crossbred bulls should be tested for dairy and beef traits, and only the absolute best bulls can be used for further breeding.

(b) Breeding programmes

The breeding programmes for the Danish cattle breeds follow the same principles as schemes in several other European countries. It means :-

1) selection of potential bull dams on EDP;
2) inspection of bull dams and insemination with semen from selected bull sires (imported semen or semen from bulls born and tested in Denmark);
3) performance tests of 600 bull calves per year and selection on an index based on growth rate and ultrasonic muscle area;
4) test insemination with approximately 120 young bulls per year;
5) registrations of the bull's calving statistics and progeny test for butterfat production and calving performance of the daughters;
6) bull sire selection : after each of the best tested bulls, 30 to 40 daughters are selected at random and inspected for milkability, temperament and exterior. Each year also 30 of the best bulls are progeny tested for beef production on a station with 10 bull calves per sire.

Future aspects

Within the next two years the evaluation of breeding value for butterfat production of cows and bulls will be based on the "Direct Updating" system.

The capacity for central performance testing will be increased and the feeding system changed to ad lib. feeding with a complete diet with a low energy concentration. Daily feed intake and feed efficiency will be included directly in the selection index.

Finally, several experiments are in progress in Denmark and in other countries to develop techniques to measure hormones, enzymes and antigen responses in growing animals. By such methods it shall hopefully in the future be possible to include an indirect selction for constitutional traits and milk production capacity in the performance test of young bulls.

References

Andersen, B.B. 1978. Animal size and efficiency with special reference to growth and feed conversion in cattle. Anim. Prod. 27, 381-391.

Andersen, B.B. 1981. Comparison of Danish Friesian, Holstein Friesian and Red Danish for Beef Production. EEC Seminar Dublin, 13-15 April 1981.

Andersen, B.B. and Andersen, G.S. 1975. Sammenhængen mellem mælkeproduktionsevne, kødproduktionsevne og kropsmal hos RDM og SDM. National Institute of Animal Science. Report No. 71, 4 pp.

Andersen, B.B., De Baerdemaeker, A., Bittante, G., Bonaiti, B. Colleau, J.J., Fimland, E., Jansen, E., Lewis, W.H.E., Politiek, R.D., Seeland, G., Teehan, T. and Werkmeister,F. 1981a. Performance Testing of Bulls in AI. Report of a Working Group of the Commission on Cattle Production. Livestock Prod. Sci. (in press).

Andersen, B.B., Jensen, J., Kousgaard, K. and Buchter, L. 1981b. Avlsstationerne for kødproduktion 1979/80. National Institute of Animal Science, Report No. 506,109pp.

Liboriussen, T. 1981. Results from 15 European Beef Breeds crossed with Danish Dairy Cows. EEC Seminar, Dublin, 13-15 April 1981.

Madsen, P. 1981. Personal communication.

Petersen, P.H., Christensen, L.G., Andersen, B.B. and Ovesen, E. 1974. Economic optimization of the breeding structure within a dual-purpose cattle population. Acta Agric.Scand. 247-259.

A NOTE ON THE COMPARISON OF HOLSTEINS AND FRIESIANS FOR GROWTH, FEED EFFICIENCY AND CARCASS TRAITS

J.R. Southgate

Meat and Livestock Commission, Queensway House, Bletchley, Milton Keynes, MK2 2EF, United Kingdom

Summary

Beef production trials comparing Holstein and Friesian cattle in the UK have shown the Holstein and the Friesian to have a similar growth rate, but slaughter weights of the Holstein have usually been greater because they have fattened more slowly. There is some evidence that the Holstein is less efficient than the Friesian.

Holstein cattle had longer carcasses and legs than Friesians and had poorer shape. This poor shape was associated with slightly inferior saleable meat yield. There was evidence of price discrimination against poor shape in UK markets.

It is estimated that the proportion of Holstein blood in 'black and white' UK cattle is increasing and a deterioration in shape in this population has been recorded.

Introduction

The Holstein only accounts for 1% of the UK dairy herd (Milk Marketing Boards, 1979). However the breed has been used widely on Friesian cows in recent years and it is estimated that 20% of the UK 'black and white' calf crop in 1981 will contain Holstein blood. Beef from the dairy herd is an important part of UK beef production (Southgate, 1981) and this level of crossbreeding with black and white cattle has serious implications. The Milk Marketing Board (1970) have evaluated Holstein x Friesians in beef production. Cook and Newton (1979) and Tas and Scott (1981) have evaluated Holsteins.

Holsteins and Friesians have been included in the MLC Breed Evaluation Programme since 1976 in 16 and 24 month dairy beef systems. In these trials, Friesian, Holstein and beef cross Friesian steers are reared on complete diets and slaughtered serially at three levels of estimated subcutaneous fat (SFe). These trials will not be complete until 1982. Interim results, for performance traits in the 16 month system are given in Table 1. Data were analysed using a model which included SFe as a regressor.

Table 1 Comparison of Holstein and Friesian steers slaughtered at 5.7% estimated sub-fat

	Friesian	Holstein
Number	45	45
Daily gain (kg)	0.84	0.83
Feed consumed (kg)	3293	4195
Slaughter weight (kg)	444	503
Slaughter age (days)	486	562
Efficiency (g gain/kg feed)	103	92

Source: MLC

Friesians and Holsteins had similar daily gain, but the Holsteins took 76 days longer to reach average SFe and were consequently 59 kg heavier. The Holsteins consumed more feed and were apparently less efficient. Southgate (1981) reports similar efficiency results for Friesians in an earlier part of the same programme, which did not include Holsteins. In these earlier results, the breeds taking longer to reach a given SFe were the least efficient and the interim results in Table 1 are similar in this respect.

There is general agreement on the lack of difference in growth rate between the two breeds (Cook & Newton, 1979 and Tas & Scott, 1981). The growth rate of Holstein x Friesians and Friesians are also similar (Milk Marketing Board, 1970). Tas and Scott (1981) also found the Holstein took longer to reach the same level of sub-fat whilst Cook and Newton (1979) in a trial involving serial slaughter at three weights, found Holsteins to have lower killing out percent, less saleable meat and more bone than Friesians.

Many of the cattle involved in the UK trials have been classified for fatness and shape by MLC staff (Table 2). The lower fatness scores for the Holstein cattle in this trial are in agreement with the other trial results concerning the time taken to reach equal fatness.

Table 2 Carcass characteristics of Holsteins relative to Friesians (Friesians = 100)

Trial	Beef system	No. of each breed	Carcass weight	MLC carcass classification	
				Fatness	Conformation
* Newcastle University	Barley	30	104	93	36
	18 month	30	104	96	45
	24 month	30	110	93	32
+ MMB/MLC	18 month	48	99	91	45
* ADAS	Barley	50	108	98	42
	18 month	40	100	88	44
* MLC	18 month	30	114	100	57
	24 month	24	107	100	40

+ End point of trial - equal weight

* End point of trial - equal fatness

Source: MLC

MLC cutting tests (Table 3) show the Holstein to have a lower proportion of saleable meat and a poorer distribution of this saleable meat than the Friesians. However, this is less than might be supposed from the shape assessment in Table 2.

The poorer shape of the Holstein, whilst being associated with a lower yield of saleable meat is also discriminated against in fatstock markets. MLC staff recorded the prices of cattle at 14 UK markets on 2 February 1981, estimating the carcass classification of each animal (Table 4).

MLC staff have also recorded a deterioration in the shape of black and white cattle in the UK (MLC Fatstock Department - personal communication). In 1975 4.6% of black and white cattle were classified 1 and Z for shape compared to 8.9% in

Table 3 Carcass composition of Holsteins and Friesians

Trial	Beef system	No. of cattle	Saleable beef* (% of carcass)	Saleable beef in higher-priced cuts (%)
Friesians				
Various	18,24 month	70	70	45
Holsteins				
MMB/MLC	18 month	30	69	44
ADAS	18 month	10	68.5	44.5
MLC	18 month	18	68.5	43.5
MLC	24 month	12	67.5	44.5

Source: MLC

1979. In view of the trial results this deterioration in shape is commensurate with an increasing level of Holstein blood in the UK black and white cattle.

Table 4 Effect of shape on the price of black and white cattle at 14 UK markets on 2 February 1981 (p/kg)

Source: MLC

	Conformation*				
Liveweight (kg)	4	3	2	1	Z
380-460	90.1 (16)	85.4 (71)	83.4 (48)	81.5 (18)	75.3 (7)
465-555	87.9 (21)	83.9 (137)	81.9 (78)	79.9 (24)	-
560 +	83.6 (22)	82.2 (134)	80.7 (84)	76.4 (17)	69.5 (5)
Overall	86.9	83.5	81.8	79.4	72.9

* shape assessment 5 - good to 1 - poor shape and Z very poor shape

References

Cook, K.N. and Newton, Jennifer, M. 1979. A comparison of Canadian Holstein and British Friesian steers for the production of beef from an 18 month grass/cereal system. Anim. Prod. 28, 41-47.

Milk Marketing Board, 1970. Report of the breeding and production division, MMB, Thames Ditton, Surrey, UK, 20, 96-97.

Milk Marketing Board, 1979. Dairy facts and figures 1979. The Confederation of United Kingdom Milk Marketing Boards, Thames Ditton, Surrey, UK 36.

Southgate, J.R. 1981. The current practice and possible future role of commercial cross-breeding for exploiting stratified resources with particular reference to the UK. EEC Seminar Dublin, Ireland, 1981 (in print).

Tas, M.V. and Scott, B.M. 1981. In Experimental Husbandry, HM Stationery Office, London (in print).

DISCUSSION ON MR. J.R. SOUTHGATE'S PAPER

Chairman: Prof. Dr. D. Smidt

Jansen: Is there any value for shape score 5 in the UK ? Your slide showed none.

Southgate: There were no cattle in this category on the day of valuation.

Jansen: Was the scoring done within breeds ?

Southgate: Yes, within Black and White cattle.

Oostendorp: Regarding the calculations shown for the value of Holstein Friesians relative to Friesians: does the final difference reflect the differences in calf prices ?

Southgate: Superficially it would appear that if one paid £20 less for the Holstein Friesians, then there was little difference between the outcomes. However, this would ignore the financial effects of the differences in age at slaughter and extra feed requirements. In continuous production systems these differences may be small, but in seasonal systems differences in maturity could have either a negative or positive effect financially, depending on the situation.

Oostendorp: So the difference in calf price needs to be added on to the differences shown ?

Southgate: Yes.

Noble: Dr. Southgate demonstrated a difference in value of about 4 pence per live kilogram between Holstein and Friesian and also showed from a market survey the influence of shape on price. He stated that the trade was discounting Holstein beef at about the right rate, but this would only be true if there was only one shape point between the breeds, when in fact it was probable that there was more than one point.

Southgate: If you are right, and there is more than one point between the breeds, then the trade is discriminating against the Holstein. However, I doubt this and it should be noted that there is considerable variation in type within the breeds and which results in a large overlap between them.

Cunningham: Southgate's figures show a net deterioration of about 5% in economic value of Holstein crosses relative to Friesians, mainly due to conformation. Flanagan's data produce a similar figure. We should try to assess the data from all countries in a similar way to see if this difference is consistent. Secondly, the range in conformation between Holstein progeny groups indicates that we could select North American sources which would do much less damage to the beef merit of our Friesians.

Southgate: I agree with the sentiments, although the opportunity for selection has been rather limited in the past, due to the lack of suitable information.

Neimann-Sorensen: Our results also show a 5% decrease in the value from using Holstein Friesians. I believe that more attention should be paid to the possibilities which a performance test and selection programme offer for all potential AI bulls, in counteracting the negative effects on beef characters by Holstein Friesian crossing. Also, selection on the basis of conformation of the bulls in the U.S.A. from which semen is imported, could be effective.

Flanagan: Our experience with Holstein Friesian bulls entering Ireland indicates that prior selection would be fruitful, as Cunningham suggests.

Neimann-Sorensen: While your results and their interpretation all appear very much in line with the Danish experience, there is one point which looks contradictory: the aptitude of the Holstein Friesian to deposit fat. You find they are less fat and we find they are more fat, especially at higher weights ?

Southgate: The difference between the two sets of results may be that fat in our experience means subcutaneous fat, while in the Danish experiments it also includes internal fat.

Langholz: Results from progeny testing in Germany support the Danish result, that Holsteins show a higher degree of internal fat deposition.

MEAT PRODUCTION OF HOLSTEIN FRIESIANS IN COMPARISON TO DUTCH FRIESIANS AND DUTCH RED AND WHITE

J.K. Oldenbroek

Institute for Animal Husbandry Research "Schoonoord," Zeist, The Netherlands.

Summary

At the moment Holstein Friesian (HF) bulls from North America are used quite extensively in the Dutch Friesian (DF) population. Some crossing with Red Holstein Friesian bulls is practised in the MRIJ population. Holstein Friesians are used in both breeds to quickly improve milk yield and udders. From a breed comparison study it can be concluded that the Holstein Friesian genes will have a negative influence on the slaughter quality of the DF animals (dressing percentage, fleshiness, percentage of saleable meat) in all three types of meat production: slaughter cows, beef bulls and veal calves. In the MRIJ population the negative influence of the Holstein Friesian genes will be more pronounced on slaughter quality and also feed conversion will be influenced unfavourably. The differences in the slaughter value of cows between the HF, DF and MRIJ population are rather small and are of minor importance, because they have to be spread over 5 years. The differences between these breeds in the value of the newborn bull calves are rather large and are of importance, because they will return on average each two years on a cow basis,

Introduction

In the last decade Holstein Friesian (HF) genes from North America were introduced in the Dutch Friesian population (DF) and on a small scale in the Dutch Red and White population (MRIJ). In the DF population 14.8 % and in the MRIJ population 1.3% of the inseminations were carried out with semen of Black and Red Holstein bulls respectively. (Annual Report AI 1979). In 1979 22.3% of the DF test bulls had a HF sire.

In the Netherlands nearly all meat from cattle is produced by animals from dual purpose breeds. In 1978 53% (on kg basis) of this meat was produced by cows and heifers, 15% by beef bulls and 32% by veal calves. In this paper the meat production of purebred Holstein Friesians, Dutch Friesians and Dutch Red and Whites in these three meat production systems will be outlined.

Table 1 Estimates of meat production characters of $(N)_2$, $(\frac{3}{4}N\frac{1}{4}A)_2$, $(\frac{1}{2}N\frac{1}{2}A)$ and $(\frac{1}{4}N\frac{3}{4}A)_2$ beef bulls (b.b.) and veal calves (v.c.) and the a, d and m effects.

subpopulation/effect	$(N)_2$	$(\frac{3}{4}N\frac{1}{4}A)_2$	$(\frac{1}{2}N\frac{1}{2}A)_2$	$(\frac{1}{4}N\frac{3}{4}A)_2$	a	d	m
Number beef bulls	32	33	28	35			
Number veal calves	32	33	32	34			
<u>Trait</u>							
Daily gain b.b. (g)	1063	1063	1070	1043	-40	27	-4
Daily gain v.c. (g)	1123	1139	1102	1126	-27	-7	27
kf St.E/kg gain b.b.	3.19	3.24	3.19	3.33	0.19	-0.09	0.05
kg milk replacer/ kg gain v.c.*	1.60	1.61	1.72	1.70	0.18	0.03	-0.05
Fleshiness b.b. $(1^- - 6^+)$*	3.6	3.3	3.0	2.5	-1.6**	0.2	-0.0
Fleshiness v.c. $(1^- - 6^+)$*	3.5	3.3	3.1	2.8	-1.0**	0.0	-0.1
Fat covering b.b. $(1^- - 6^+)$*	3.5	3.4	3.4	3.1	-0.6**	0.3**	-0.1
Fat covering v.c. $(1^- - 6^+)$	3.1	3.0	2.8	3.0	0.0	-0.3	-0.1
Dressing percentage b.b.*	59.2	58.4	58.2	57.9	-0.8	-0.5	-0.4
Dressing percentage v.c.*	64.1	64.2	63.5	62.9	-2.5**	0.6	0.4*

* Significant differences between the offspring of Holstein Friesian and Dutch Friesian bulls (P 0.05)
** Significant effect ($P < 0.05$)

These HF, DF and MRIJ animals participated in a breed comparison described by Oldenbroek (1974). But firstly, the meat production data of two crossbreeding trials with Holstein Friesians and Dutch Friesians will be summarised.

Politiek c.s. (1977) fattened purebred Dutch Friesians and animals with 75% Holstein Friesian genes as veal calves for 19 weeks. The 75 % HF animals had a 1.2% lower dressing percentage and a lower score for fleshiness compared to the DF animal. No significant differences between these groups were found for growth and feed conversion.

Oldenbroek (1980) found in a crossbreeding trial with DF, 25% HF-, 50% HF- and 75% HF animals (respectively $(N)_2$, $(\frac{3}{4}N\frac{1}{4}A)_2$, $(\frac{1}{2}N\frac{1}{2}A)_2$ and $(\frac{1}{4}N\frac{3}{4}A)_2$) no heterotic (d) and maternal (m) effects of importance in beef- and veal production. Only for slaughter quality characters significant breed (a) effects were calculated, as can be seen in Table 1.

Meat production from slaughter cows

In the comparison of HF, DF and MRIJ live weight, slaughter weight and classification was known of 92, 132 and 141 cows (age $>$ 700 days) respectively. They were slaughtered in the period 1973-1980. The main data are summarised in Table 2.

Table 2 Meat production data of HF, DF and MRIJ slaughter cows

Breed	HF	DF	MRIJ
Number	92	141	132
Live weight (kg)	634.9	573.9	590.0
Slaughter weight (kg)	304.8	280.3	296.8
Dressing percentage**	47.8	48.7	50.2
Classification (1= best, -18= poor)**	13.8	11.2	8.9
Age at slaughter (days)	1819	1704	1823
Corrected slaughter weight (kg)*	304.9	284.9	296.8
Price per kg* (f)	5.72	5.98	6.21
Slaughter value (f)	1744	1704	1843

* corrected, see text
** $p < 0.05$ (significant differences between breeds)

Due to a high number of calves born alive and reared in the DF group, more animals could be culled in this group at a young age. Therefore the slaughter weight of the DF (and HF) animals (with linear regression within breeds) was corrected to an age of 1823 days. The price per kg is calculated with the average classification results and the price level and -differences of the first week of January 1981. Due to a higher live weight at slaughter the corrected slaughter weight of the HF animals is 20 kg higher than that of the DF animals. Despite a lower classification the slaughter value of a HF cow is on average f 40.- higher than that of a DF cow. Despite a somewhat higher corrected slaughter weight but due to a much lower classification the slaughter value of a HF cow is f 99.- lower than that of a MRIJ cow.

Meat production from beef bulls

In 1975-1976, 27 HF, 34 DF and 28 MRIJ bull calves were fattened as beef bulls with corn silage and (a protein rich) concentrate. The animals were slaughtered at 420, 460 or 500 kg live weight. After slaughter the right half of the carcass was divided into saleable meat, fat and bone, according to a commercial dissection method described by Bergström c.s. (1967). For none of the characters could a significant breed x slaughter weight interaction be calculated. Therefore the results in Table 3 are presented at the average slaughter weight of 460 kg.

HF beef bulls had a poorer feed conversion, a lower score for fleshiness, a lower percentage of saleable meat and a higher percentage of bone than DF and MRIJ beef bulls. Therefore their carcass value was f 81 and f 118 lower, and their feed costs were f 19 and f 67 higher in comparison with DF and MRIJ beef bulls respectively.

Meat production from veal calves

In 1976 23 HF, 38 DF and 28 MRIJ calves were fattened with milk replacer up to 190 kg live weight. The results are summarised in Table 4.

Table 3 Meat production data of HF, DF and MRY beef bulls

Breed	HF	DF	MRY
Number	27	34	28
Age at slaughter (days)	397	399	393
Daily gain (g)	1062	1065	1072
Feed conversion (kg St.E/kg gain)**	3.26	3.17	2.98
Dressing percentage**	57.1	58.9	58.8
Fleshiness (1= poor -6= best)**	2.06	3.35	3.54
Fat covering (1= poor -6= fat)**	3.05	3.51	2.85
Saleable meat (% in carcass)**	65.2	66.3	67.0
Fat (% in carcass)**	19.0	19.4	18.2
Bone (% in carcass)	15.8	14.2	14.8
Carcass value (f)*	1772	1853	1890
Feed costs (f)	955	936	888
Value calf (f)	0	+100	+185

* calculated after dissection
** $p < 0.05$ (significant differences between breeds)

Table 4 Meat production data of HF, DF and MRIJ veal calves

Breed	HF	DF	MRIJ
Number	23	38	28
Age at slaughger (days)*	126	131	125
Daily gain (g)*	1119	1128	1161
Kg feed/kg gain*	1.69	1.59	1.54
Dressing percentage*	61.1	63.1	63.3
Fleshiness (1= poor -6= best)*	2.3	3.5	3.8
Fat covering (1= poor -6= fat)*	2.6	3.1	3.2
Meat colour (1= light -6= dark)	1.6	1.7	1.6
Carcass value (f)	951	1040	1065
Feed costs (f)	449	445	423
Value calf (f)	0	+93	+140

*$p < 0.05$ (significant differences between breeds)

Price differences between veal calves are mainly determined by differences in fleshiness and meat colour (Laurijsen c.s. 1978). In this trial the differences in carcass value are due to differences between the breeds in fleshiness. The HF veal calves had a poorer feed conversion, a lower dressing percentage and a lower score for fleshiness and fat covering in comparison to the DF and MRIJ veal calves. The carcass value of the HF veal calves was f 89 and f 114 lower and the feed costs were f 4 and f 26 higher in comparison with the DF and MRIJ veal calves respectively.

Discussion

Holstein Friesian genes have been introduced in the Dutch Friesian population and in the MRIJ population to improve quickly milk yield and udders. For the Dutch dairy farmer, at least 75% of his income comes from milk sales. Therefore, the need for high producing cows with a long herd life is stressed and therefore his use of Holstein Friesian genes is justified. But from all the data presented in this paper, it will be clear that the u se of Holstein Friesian will have a large negative influence on the slaughter quality of the DF animals (dressing percentage, fleshiness, percentage of saleable meat).

Differences in value between slaughger cows are much less important than differences in value between young bull calves. The former have to be spread over approximately 5 years and the latter will return on average each two years on a cow basis.

In the Netherlands all the young male calves are sold within 10 days to specialised fatteners. The price is determined by breed (=hair colour), weight and type. There is a close positive relationship between weight and type within breeds (Laurijsen c.s. 1980). Within its weight class a HF calf is scored low for type, but within a weight class it should be scored much lower. As Holstein Friesian calves are approximately 6 kg heavier than Dutch Friesian calves it will be clear that the fattener will pay too much for a calf with HF genes and will therefore earn more money with a DF calf than with a calf with HF genes.

In a second way the introduction of Holstein Friesian genes

has a negative influence on the meat production of a herd. With the use of HF bulls the percentage of calving difficulties increases (Oldenbroek, 1980) and Holstein Friesians have more calves born dead compared to Dutch Friesians (Oldenbroek, 1979).

It will be clear, that the Holstein Friesian will have a more pronounced influence on the meat production capacity of the MRIJ breed (an unfavourable feed conversion and a lower slaughter quality). Therefore, the optimal amount of HF genes in the MRIJ population will be much less than in the DF population.

At the moment it is possible to minimise the negative influence of the HF breed on the slaughter quality of the DF population by buying HF bulls with a better fleshiness than the average HF bull and by paying attention to fleshiness in the performance test of bulls.

In the MRIJ population some Red Holstein Friesian crossbred bulls (RHF x MRIJ) are tested at the moment.

On the way to a more specialised dairy breed some crossing with beef breeds could be justified in the DF population. But up to now, crossbreeding with beef breeds is only practised in Holland on a small scale (less than 1% of the total amount of inseminations in 1979) because the dairy farmer does not accept more calving difficulties.

References

Bergström, P.L. and Kroeske, D. (1967). Methods of carcass assessment in research on carcass quality in the Neterlands (I and II). Report C-123, Institute for Animal Husbandry Research "Schoonoord", Zeist.

Laurijsen, H.A.J., Nijeboer, H., De Boer, Tj. and Jansen, A.A.M. Appendix (1978). The application of the IVO classification system in classifying fattened calves. Report C-332 (Dutch). Institute for Animal Husbandry Research "Schoonoord",Zeist.

Laurijsen, H.A.J., Van Eldik, P. and Minkema, D. (1980). Classification of new born calves as a predictor of suitability for beef production. Report B-163 (Dutch, English summary). Institute for Animal Husbandry Research "Schoonoord," Zeist.

Oldenbroek, J.K. (1979). Comparison of Holstein Friesian-, Dutch Friesian and Dutch Red and White cattle. De Friese Veefokkerij (1979) : 3 : 152-157 (Dutch).

Oldenbroek, J.K. (1980). Breed and crossbreeding effects in a crossing experiment between Dutch Friesian and Holstein Friesian cattle. Livestock Production Science 7(1980) 3 : 235-241.

Politiek, R.D., Vos, H., Brandsma, H.A., De Leede, C.A. and Van Wolfswinkel M. (1979). Veal calves sired by Dutch-American- and British-Friesian bulls. Bedrijfsontwikkeling (8) 1977, 914-918 (Dutch).

ABILITY OF HOLSTEIN-FRIESIAN, DUTCH BLACK AND WHITE AND BELGIAN WHITE-RED BULL CALVES FOR VEAL AND BEEF PRODUCTION

D.L. De Brabander, Ch.V. Boucque, M. Casteels,
J.V. Aerts and F.X. Buysse

National Institute for Animal Nutrition,
Scheldeweg 68, B-9231 Melle-Gontrode, Belgium.

Summary

The ability of Holstein-Friesian, Dutch Black and White and Belgian White-Red males for beef production was studied in three trials. In the first trial 21, 22 and 20 male calves of each breed respectively, were compared for veal production and meat quality aspects. In the second trial, 17, 18 and 54 calves were used to study their ability for intensive beef production (final liveweight ± 470 kg), whereas in the third trial, 12, 5 and 20 calves were semi-intensively fattened until a final liveweight of ± 550 kg. Carcass composition and meat quality properties were also examined.

In all beef production systems, the Holstein breed had the lowest growth rate, the highest feed intake per kg live weight gain and consequently the highest feed cost price per kg gain. Moreover, they were sold at the lowest price. Also dressing percentage and meat production coefficient (only determined for the intensive and semi-intensive beef production systems) were the most unfavourable for the Holstein breed.

Except for the colour of the veal calves, the meat quality characteristics were not significantly different between the breeds. In all systems, the best results were obtained with the Belgian White-Red breed. Differences between the breeds were lowest in veal production.

Introduction

In Belgium, beef production can be interpreted as a by-product of the dairy herd, since on a total of 1,113,271 cows in 1980, there were 976,032 (=87.7%) dairy cows and 137,239 (=12.3%) suckler cows (Census 15.05, 1980 :I.N.S.). About 46% of the Belgian cows belong to dairy breeds, 42% to dual purpose breeds and 12% to beef breeds. The male animals slaughtered in 1979 can be divided as follows : 11% steers, 42% bulls, and 47% veal calves (I.N.S. 1980). Bull fattening and veal are important beef production goals in Belgium.

Since 1971, the dairy herd at our Institute consisted of

purebred cows belonging to the Holstein-Friesian, the Dutch Black and White and the Belgian White-Red breed. To study the ability of these breeds for beef production, the male descendants were intensively and semi-intensively fattened under similar conditions to a final liveweight of $\pm$ 470 and $\pm$ 550 kg respectively. In 1980 these three breeds were compared for veal production.

Experimental design

Veal calves

All calves were purebred and were bought on practical farms. Because it was impossible to have the desired number of calves on the same date, the trial started on three dates, with an interval of 2 weeks between each series. Calves which had too low initial hematocrite level (3.8%) received once 1 g $FeSO_4$. They were only fed milk replacer and water. The milk replacer fed up to the age of 6 weeks contained 22% crude protein, 25% sugar + starch, and 18% fat. After 6 weeks the crude protein, sugar + starch and fat levels were 21, 26 and 19% respectively. Feeding level was individually adapted to appetite, but was reduced when digestive upsets occurred. The calves were sold when the desired degree of fatness was attained. The meat colour, an important commercial factor, was determined 24 hr after slaughtering on samples of the Longissimus dorsi muscle. The visual colour was objectively determined (Hunter and Gofo); the pigment content was determined by extraction methods. As daily gain seemed to be influenced by the series number (season) and because the number of calves differed for the 3 breeds in each series, the method of fitting constants was applied for the statistical analysis of the results.

Intensive beef production

All male descendants of our dairy herd which were born in the period January-June of each year (1972-1977) were intensively and individually fattened to a live weight of $\pm$ 470 kg (baby-beef type). In the rearing period, which was finished in the 17th week, they were offered fresh milk for 3 weeks, milk replacer until the 10th week, calf starter restricted to 3 kg a day and grass hay ad libitum. Between the 17th week and the 7th month, the diet consisted of grass hay ad libitum and 3 kg of a concentrate (± 62 starch equivalent; ± 10% DCP) that contained 70% dried sugar beet pulp. The fattening period started in the 7th month. The same concentrate, as well as straw, were voluntarily administered.

Semi-intensive beef production

The male calves born from July to December were each year (1972-1976) fattened according to a semi-intensive method. The diets in the rearing period, as well as in the period between the 17th week and the beginning of the pasture period, were similar to those for the intensively fattened bulls. From April, calves which had attained an age of 6 months went to pasture and received a supplement of 2 kg dried sugar beet pulp daily. After the pasture period, they were intensively fattened, using the same diet as in the intensive beef production system. The results of both bull fattening trials were statistically analysed by means of the one-way analysis of variance.

The meat quality characteristics for both production systems were determined approximately 24 hr post-mortem on the Longissimus dorsi muscle (6th rib). The visual colour, the pigment-content and the water holding capacity were studied on all animals involved.

Results and discussion

1. Veal calves

The most important beef production results, as well as the meat quality characteristics are mentioned in Tables 1 and 2 respectively.

The mean total liveweight (LW) gain was nearly the same for each breed. Daily LW gain for the whole period was significantly lower for the Holsteins than for the White-Red calves, while the Dutch Black and White were intermediate in that respect. Dijkstra et al. (1975) also obtained a lower daily gain with Holstein veal calves than with Dutch White and Red, and Dutch Black and White calves. Differences in daily gain were most pronounced in the second part of the experiment (Table 1, Figure 1). The quantity of milk replacer consumed per kg live

Table 1 Beef production results of veal calves ($\pm s_{\bar{x}}$)

Breed	HF [1]	Bl.Wh.	Wh.R.
Number	21	22	20
Initial weight (kg)	44.4	41.5	46.5
Final weight (kg)	182.9	180.4	186.4
Liveweight gain (kg)	138.5	138.9	139.9
Experimental days	139.8	133.5	132.9
Intake milk replacer (kg)			
Total	252.0	237.5	234.7
Per day	1.80	1.78	1.77
Daily liveweight gain (g)			
0 - 8 weeks	675^{a} [2] $\pm$ 17	695^{a} $\pm$ 17	703^{a} $\pm$ 18
9 w - slaughtering	1224^{a} $\pm$ 26	1282^{ab} $\pm$ 25	1308^{b} $\pm$ 26
Total period	1000^{a} $\pm$ 17	1036^{ab} $\pm$ 17	1053^{b} $\pm$ 18
Milk replacer/kg gain (kg)			
0 - 8 weeks	1.55^{a} $\pm$ 0.03	1.51^{a} $\pm$ 0.03	1.49^{a} $\pm$ 0.03
9 w - slaughtering	1.92^{a} $\pm$ 0.03	1.82^{b} $\pm$ 0.03	1.78^{b} $\pm$ 0.03
Total period	1.81^{a} $\pm$ 0.03	1.73^{b} $\pm$ 0.03	1.69^{b} $\pm$ 0.03
Dressing percentage	61.6^{a} $\pm$ 0.3	62.9^{b} $\pm$ 0.3	64.8^{c} $\pm$ 0.3
Feed cost price (BF)			
Per kg liveweight gain	53.3^{a} $\pm$ 0.8	50.9^{b} $\pm$ 0.8	49.7^{b} $\pm$ 0.8
Selling price (BF)			
Per kg liveweight	51.6^{a} $\pm$ 0.7	59.6^{b} $\pm$ 0.7	62.8^{c} $\pm$ 0.9
Per kg carcass weight	83.6^{a} $\pm$ 1.0	94.8^{b} $\pm$ 1.0	96.8^{b} $\pm$ 0.9

(1) HF = Holstein-Friesian, Bl.Wh. = Dutch Black and White, Wh.R. = Belgian White-Red.

(2) Means on a same line bearing the same superscript letter are not significantly different ($P > 0.05$)

FIGURE 1 Liveweight evolution of veal calves

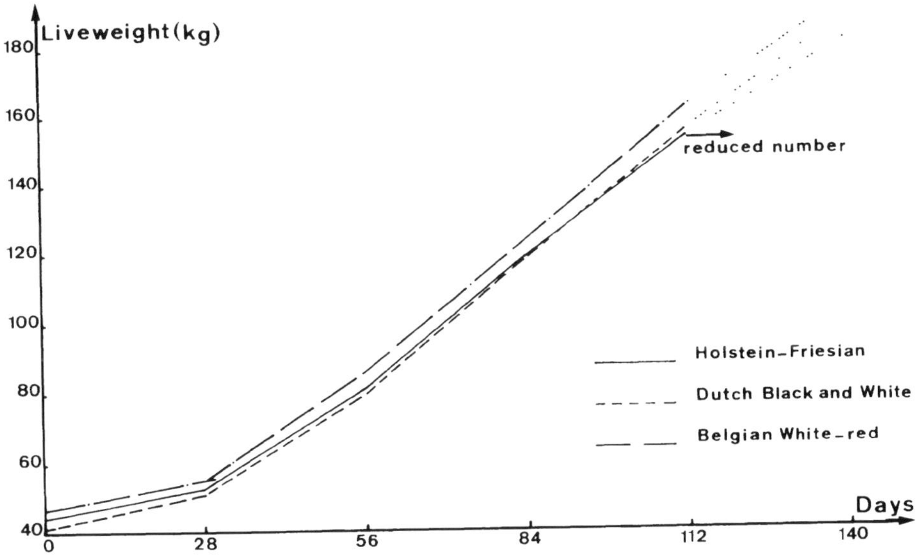

2. Intensive beef production

The most important results reflecting beef production ability of the intensively fattened bulls are represented in Tables 3 and 4.

Growth rate during the rearing period was slightly higher for the Holstein calves.

The fattening period lasted 265, 262 and 242 days respectively. Although the White-Red bulls were slaughtered at a markedly higher final live weight, they had a significantly lower fat content in the carcass (Table 4). The Holstein bulls as well the Dutch Black and White ones had a significantly lower daily gain than the Belgian White-Red breed. The latter had a higher daily feed intake. The lower growth rate, as well as the higher degree of fatness of the first two breeds, resulted in a significantly higher feed intake and feed cost price per kg LW gain. Also for the total period, the White-Red breed was superior. Oldenbroek (1976) and De Boer et al. (1976) also obtained a higher growth rate and a better feed conversion with Dutch White and Red bulls than with Holstein or Dutch Black and White bulls.

weight gain was significantly higher for the Holstein calves compared with the other two breeds, which is in agreement with the results of Oldenbroek (1976) and Laurijsen and Oldenbroek (1978). Also in that respect, the second part of the fattening period was mainly responsible for these differences. Dressing percentage differed significantly between the three breeds and was the lowest for the Holstein breed. The lower feed efficiency for the Holsteins resulted in a higher feed cost price per kg LW gain. On the other hand, the selling price per kg LW of the Holstein calves was 8 BF and 11.2 BF lower compared to the Dutch Black and White and the Belgian White-Red calves respectively. Based on these results and admitting that the production costs per calf, except feed costs, are equal for the three breeds, a newborn male Holstein calf for veal should have a value of ± 1,700 and ± 2,700 BF lower than the Dutch Black and White and Belgian White-Red calves respectively.

The results generally indicate that veal calves produced in the same circumstances as in this experiment are not economically attractive.

The data shown in Table 2 indicate that there was no significant difference between the blood parameters (Haemoglobin and Haematocrite) of the three breeds.

Table 2 Meat quality characteristics and blood parameters of veal calves ($\pm s_{\bar{x}}$)

	HF		Bl.Wh.		Wh.R.	
Number	21		22		20	
Initial Haemogl.cont. (g/100 ml)	9.92	± 0.36	10.98	± 0.40	10.52	± 0.54
Final Haemogl.cont. (g/100 ml)	9.15	± 0.32	9.28	± 0.27	8.64	± 0.29
Initial Haematocr.cont.(%)	41.1	± 1.3	45.3	± 1.9	42.7	± 2.6
Final Haematocr.cont.(%)	35.8	± 1.3	37.4	± 0.8	34.7	± 1.0
Hunter						
L (lightness)	47.70[ab]	± 0.61	46.19[b]	± 0.64	48.16[a]	± 0.49
a) chromacity	9.41	± 0.21	10.30	± 0.23	9.75	± 0.31
b) chromacity	8.50	± 0.15	8.13	± 0.20	8.51	± 0.17

Table 3 Beef production results of intensively fattened bulls ($\pm s_{\bar{x}}$)

Breed	HF	Bl.Wh.	Wh.R.
Number	17	18	54
Rearing period			
Initial weight (kg)	41.5	35.6	47.2
Final weight (kg)	124.6	115.8	128.2
Experimental days	113.5	113.9	114.6
Daily liveweight gain (g)	$732^{a} \pm 32$	$706^{a} \pm 30$	$706^{a} \pm 15$
Total feed intake (kg)			
- milk	58.8	59.5	66.7
- milk replacer	27.1	27.0	26.5
- calf starter	169.0	152.7	158.1
- grass hay	37.1	28.4	29.4
Fattening period			
Initial age (days)	202	188	194
Final age (days)	467	450	436
Initial weight (kg)	191	179	200
Final weight (kg)	460	452	494
Daily liveweight gain (g)	$1022^{a} \pm 25$	$1043^{a} \pm 20$	$1223^{b} \pm 16$
Daily intake			
- concentrate (kg)	$7.73^{ab} \pm 0.18$	$7.70^{a} \pm 0.11$	$8.04^{b} \pm 0.08$
- starch units (kg)	4.87	4.84	5.01
- dig.crude protein (g)	764	767	805
Intake/kg LW gain (kg)			
- concentrate	$7.58^{a} \pm 0.19$	$7.38^{a} \pm 0.15$	$6.60^{b} \pm 0.08$
- starch units	$4.78^{a} \pm 0.12$	$4.66^{a} \pm 0.12$	$4.12^{b} \pm 0.05$
Feed cost price/kg LW gain (BF)	$53.4^{a} \pm 1.3$	$52.8^{a} \pm 1.1$	$46.4^{b} \pm 0.6$
Total period			
Daily liveweight gain (g)	$898^{a} \pm 14$	$926^{a} \pm 14$	$1030^{b} \pm 11$
Feed cost price/kg LW gain (BF)	46.4	45.4	41.3
Selling price (BF) (x)			
- per kg LW	$41.5^{a} \pm 1.3$	$56.3^{b} \pm 1.0$	$62.1^{c} \pm 0.4$
- per kg carcass weight	$71.9^{a} \pm 2.0$	$94.8^{b} \pm 1.6$	$102.9^{c} \pm 0.6$

x Selling price adjusted to a price level of 1980

Table 4 Carcass and meat quality of intensively fattened bulls ($\pm s_{\bar{x}}$)

Breed	HF	Bl.Wh.	Wh.R.
Number	17	18	54
Dressing percentage	$57.7^a \pm 0.4$	$59.3^b \pm 0.3$	$60.2^b \pm 0.3$
Carcass composition (%)			
- meat	$58.7^a \pm 0.5$	$60.2^a \pm 0.7$	$64.3^b \pm 0.4$
- fat	$25.2^a \pm 0.6$	$26.0^a \pm 0.8$	$21.4^b \pm 0.5$
- bones	$16.1^a \pm 0.3$	$13.7^b \pm 0.3$	$14.3^b \pm 0.2$
Meat production coeff. (%)	$33.8^a \pm 0.3$	$35.7^b \pm 0.4$	$38.7^c \pm 0.3$
Meat quality characteristics			
Colour			
- Reflection (Göfo)	87.3 ± 1.0	85.5 ± 0.9	84.9 ± 0.9
- Hunter			
L (Lightness)	33.70 ± 0.46	33.64 ± 0.52	34.06 ± 0.34
a) chromacity	15.78 ± 0.40	15.78 ± 0.45	15.81 ± 0.25
b) chromacity	8.20 ± 0.20	8.19 ± 0.20	8.33 ± 0.16
Pigment content			
- Hornsey (ppm haem.)	$167.6^a \pm 10.0$	$191.7^a \pm 9.1$	$174.8^a \pm 6.0$
- Extraction (Extinction)	$0.308^a \pm 0.013$	$0.287^a \pm 0.012$	$0.258^a \pm 0.006$
Waterholding capcity (cm^2)	3.5 ± 0.2	3.5 ± 0.2	3.7 ± 0.1

There was a big difference in selling price per kg LW or per kg carcass weight between the breeds. Even if selling price is expressed per kg meat, the Holsteins obtained the most unfavourable price. The latter can mainly be explained by a lack of interest for such a type of carcass on the Belgian retail market. For the Holsteins, feed cost price per kg gain even exceeded the selling price.

As with the veal calves, dressing percentage was the lowest for the Holstein breed and the highest for the White-Red breed (Table 4). Carcass composition (Martin and Torreele, 1962) as well as the meat production coefficient, representing the quantity of meat per 100 kg fasted LW, were obviously in favour of the White-Red breed.

With regard to the meat quality characteristics (visual colour, pigment content and water holding capacity), no significant differences could be observed between the three breeds.

3. Semi-intensive beef production

The number of semi-intensively fattened Dutch Black and White bulls was only 5, so that results of that breed must be interpreted carefully.

Growth rate during the rearing period (Table 5) was rather similar to that for the intensively fattened bulls (Table 3). Here again, the Holsteins had, in accordance with the higher calf starter intake, a somewhat higher weight gain in that period. The higher concentrate intake of Holstein calves during the rearing period is in agreement with the results of Van Opstal and Oldenbroek (1978), who compared Holstein-Friesians with Dutch Friesians and Dutch Red and Whites.

The fattening period lasted 237, 186 and 225 days respectively. Also for this beef production system, the White-Red breed attained a higher final live weight. Although differences in daily gain were smaller than in the previous beef production system, again the Holsteins and the White-Red breed had the lowest and the highest daily gain respectively. Feed intake and feed cost price per kg LW gain were the lowest for the White-Red bulls. Also on the total period, daily live weight gain and feed cost price were in favour of the White-Red breed.

Table 5 Beef production results of semi-intensively fattened bulls ($\pm s_{\bar{x}}$)

Breed	HF	Bl.Wh.	Wh.R
Number	12	5	20
Rearing period			
Initial weight (kg)	40.4	37.0	43.5
Final weight (kg)	124.5	119.4	122.6
Experimental days	114.9	113.6	113.8
Daily liveweight gain (g)	$731^{a} \pm 22$	$724^{a} \pm 27$	$694^{a} \pm 27$
Total feed intake (kg)			
- milk	59.4	57.5	60.5
- milk replacer	27.0	27.4	27.1
- calf starter	165.7	157.8	148.8
- grass hay	26.8	33.1	25.3
Fattening period			
Initial age (days)	379	371	354
Final age (days)	616	557	579
Initial weight (kg)	288	301	288
Final weight (kg)	556	527	571
Daily liveweight gain (g)	$1146^{a} \pm 53$	$1217^{ab} \pm 116$	$1268^{b} \pm 32$
Daily intake			
- concentrate (kg)	$9.64^{a} \pm 0.32$	$10.33^{a} \pm 0.44$	$9.73^{a} \pm 0.18$
- starch units (kg)	6.03	6.42	6.02
- dig. crude protein (g)	984	1023	965
Intake/kg LW gain (kg)			
- concentrate	$8.50^{ab} \pm 0.29$	$8.68^{a} \pm 0.56$	$7.72^{b} \pm 0.17$
- starch units	$5.32^{a} \pm 0.18$	$5.39^{a} \pm 0.33$	$4.77^{b} \pm 0.10$
Feed cost price/kg LW gain (BF)	$59.5^{ab} \pm 2.0$	$60.7^{a} \pm 3.9$	$54.1^{b} \pm 1.2$
Total period			
Daily liveweight gain (g)	$842^{a} \pm 22$	$878^{ab} \pm 19$	$913^{b} \pm 14$
Feed cost price/kg LW gain (BF)	48.3	45.0	44.3
Selling price (BF) (x)			
- per kg LW	$48.4^{a} \pm 1.0$	$57.8^{b} \pm 1.6$	$63.9^{c} \pm 0.7$
- per kg carcass weight	$83.4^{a} \pm 1.9$	$95.8^{b} \pm 2.9$	$104.4^{c} \pm 1.5$

x selling price adjusted to a price level of 1980

Compared with the results of the fattening period of the intensive system (Table 3), daily growth rate and feed intake were higher in the present system (phenomenon of compensatory growth). Feed intake and feed cost price per kg LW gain, however, were less favourable (higher liveweight interval). On the total period, however, this higher feed cost price per kg LW gain of the fattening period, has partly been compensated for by the cheaper pasture period in this system.

A big difference in selling price between the two extreme breeds is demonstrated, either when the price is expressed per kg carcass, or per kg fasted live weight. The selling price of the Holsteins only equalled feed cost price per kg LW gain.

This trial confirmed that the White-Red breed has the highest dressing percentage, the highest meat content and consequently a markedly higher meat production coefficient than the other two breeds. (Table 6).

Conclusions

- The results of the three experiments demonstrate that the Holstein male calves are markedly less suitable for beef production than the other two breeds. Differences were the lowest in veal production.
- Because a Holstein type is not desired on the Belgian beef market, their selling price is very low.
- Fattening Holstein bulls in circumstances such as our intensive and semi-intensive beef production system is not at all profitable in Belgium.
- In all systems, the best results were obtained with the Belgian White-Red breed.
- Except for the visual colour score of the Dutch Black and White veal calves (darker), no significant differences could be established for the meat quality parameters between the three breeds.

Acknowledgement

The authors are indebted to ir. R. Moermans (Biometric Unit of the Centre for Agricultural Research at Merelbeke)for the statistical analysis and to irs. R. Verbeke and G. Van de Voorde

TABLE 6 Carcass and meat quality of semi-intensively fattened bulls ($\pm s_{\bar{x}}$)

Breed	HF	Bl.W.	Wh.R.
Number	12	5	20
Dressing percentage	$58.1^a \pm 0.4$	$60.4^b \pm 0.5$	$61.3^b \pm 0.5$
Carcass composition (%)			
- meat	$59.6^a \pm 0.8$	$58.8^a \pm 1.6$	$63.1^b \pm 0.9$
- fat	$25.6^a \pm 0.9$	$28.2^b \pm 2.0$	$23.7^a \pm 0.9$
- bones	$14.8^a \pm 0.3$	$13.0^b \pm 0.5$	$13.2^b \pm 0.3$
Meat production coeff. (%)	$34.6^a \pm 0.5$	$35.5^a \pm 0.9$	$38.7^b \pm 0.7$
Meat quality characteristics			
Colour			
- Reflection (Göfo)	$84.68^a \pm 1.02$	$86.50^a \pm 1.98$	$82.30^a \pm 1.21$
- Hunter			
L (Lightness)	$33.62^a \pm 0.73$	$32.66^a \pm 0.71$	$34.17^a \pm 0.60$
a) chromacity	$16.11^a \pm 0.54$	$17.03^a \pm 1.53$	$16.47^a \pm 0.35$
b) chromacity	$8.14^a \pm 0.23$	$8.62^a \pm 0.82$	$8.53^a \pm 0.24$
Pigment content			
- Hornsey (ppm haem.)	197.0 ± 8.9	191.8 ± 11.8	184.6 ± 11.7
- Extraction (Extinction)	0.303 ± 0.015	0.288 ± 0.021	0.299 ± 0.013
Waterholding capacity (cm^2)	3.6 ± 0.1	3.4 ± 0.6	3.6 ± 0.2

(Study Centre Quality Improvement of Beef at Melle) for the determination of the carcass composition. They also gratefully thank Mr. A. Opstal, M. Desmet (Technical Engineer), A. Hertegonne, and Mrs. L. Goudepenne and M. De Bruycker for skilled technical assistance.

References

De Boer, Tj., Oldenbroek, J.K., Laurijsen, H.A.J., De Jongh, G. and Bergström, P.L. 1976. Vergelijkende mestproef met Noordamerikaanse zwartbonte (HF) en Nederlandse zwartbonte (FH) en Nederlandse roodbonte (MRY) vleesstieren. Rapport C-296 IVO "Schoonoord" (Nederland).

Dijkstra, M., Oldenbroek J.K. and Bergstrom, P.L. 1975. De geschiktheid van HF, FH, MRIJ stierkalveren voor de kalfsvleesproduktie. IVO rapport C-268.

I.N.S. 1980. Statistiques agricoles. No. 5-6.

Laurijsen, H.A.J. and Oldenbroek, J.K. 1978. De geschiktheid voor kalfsvleesproduktie van FH, HF en MRY-stierkalveren. Bedrijfsontwikkeling 9 : 879.

Martin, J. and Torreele, G. 1962. L'appréciation de la qualité des carcasses bovines par la découpe du morceau tricostal. 7,8,9. Ann. Zootech. 11 : 217.

Oldenbroek, J.K. 1976. Vergelijking van Holstein Friesians, Nederlandse zwartbonten en Nederlandse roodbonten II. De Keurstamboeker 58 : 740.

Van Opstal, A.J.C. and Oldenbroek, J.K. 1978. Vergelijking van de voederonname van Holstein-Friesian, Nederlandse zwartbonte en Nederlandse roodbonte kalveren tijdens de opfokperiode. Rapport C-350 IVO "Schoonoord" (Nederland).

DISCUSSION ON IR. J.K. OLDENBROEK'S & MR. Ch.V. BOUCQUE'S PAPERS

Chairman: Prof. Dr. D. Smidt

Langholz: Both trials were terminated at a fixed final weight. The termination point in experiments of this nature is very important to interpretation. If feeding is to the same weight for age, results can be misleading. For correct economic interpretation the termination point needs to be at the same degree of fatness. The value of serial slaughtering is obvious if the objective is to define the optimum degree of fatness.

Oldenbroek: I would like to comment. The results of beef bulls which I have just presented came from our second experiment. In the first experiment we slaughtered the animals at the same degree of finish (estimated subcutaneous fat), but after dissection the Holstein Friesian bulls had the highest percentage of fat in the carcass and the MRY bulls the lowest. The distribution of fat in the carcass varied between breeds. This makes it very difficult to slaughter animals of different breeds at the same degree of fatness.

In the experiment with beef bulls I presented, we slaughtered the animals at three different weights, 420, 460 and 500 kg. We found no interactions between breed and slaughter weight for any characters. For an economic comparison of beef bulls, our experience is that it is of no consequence whether slaughter is at constant weight or constant degree of fat cover. When slaughtering at constant weight, the higher profit of the MRY comes partly from a higher amount of meat in the carcass and partly from lower feed costs. When slaughtered at the same degree of fat cover, the higher profit of the MRY comes from a higher amount of meat per animal.

Liboriussen: Our results indicate that net profit per year will depend markedly on slaughter weight, and that in general there is no weight by breed interaction.

Southgate: Liboriussen's comments on the relationship between breeds, slaughter weight and net profit do not include the effect of seasonality. There may be situations where the seasonal limitations are of greater importance in determining net profit levels that are of overriding importance.

Oostendorp: What has been the influence of the negative results with Holstein crosses in veal and bull beef production on the prices of newly born calves in Belgium ?

Boucque: The actual price of a new born bull calf of purebred Holstein Friesian reaches only 1500 Belgian francs, compared to 6000 Belgian francs for a White Red dual-purpose calf. Even with this price difference, it is not always possible to make a profit by fattening the Holstein Friesian. Only when a Holstein Friesian calf can be sold for breeding purposes, can a higher price be obtained.

CHAIRMAN'S COMMENTS AT CLOSE OF SESSION I

Chairman : Prof. D. Smidt

I have got the impression from the papers presented in this session that in the various EEC member states, the suitability of dairy breeds and strains has been adequately analysed. This provides a more or less reliable source of valid statements and judgements on the respective national situations.

With respect to generalisation, beef characteristics and beef performance show in principle similar differences among the various "biological types" of dairy and dual-purpose cattle. So, some kind of general characterisation of suitability for beef production might be gained from the results reported here, provided the term "biological types" is used in a rather broad sense.

Of course, detailed knowledge on the large number of "breeds and strains" within dairy and dual purpose populations has been and still is accumulated. These results, however, are often valid only for the special conditions of respective trials. Complete standardisation of these conditions throughout Europe can hardly be achieved. Therefore, in order to develop national, regional or local perspectives of breeding and production policies, more attention has to be paid to the respective differences in structure, production systems, resources and overall economic conditions. This, again, requires different interpretations concerning the suitability of dairy breeds and strains for beef production.

Thus, partial genetic insufficiencies may be overcome by employment of properly adjusted production and management systems.

In this light, international exchange of results and experience is necessary and valuable.

RELATIVE PERFORMANCE OF BEEF BREEDS AND THEIR DAIRY CROSSES IN IRELAND

T.J. Teehan

Department of Agriculture, Kildare Street, Dublin 2, Ireland

Summary

The origin and use of the traditional Hereford and Angus breeds and of the newer Continental (Charolais, Simmental, Limousin, Blonde d'Aquitaine and Belgian Blue) breeds is outlined. Comparisons of pure breed performance based on on-farm records and performance test station data are presented. Comparisons between the different beef and dairy crossbreds based on beef progeny test data are also presented.

Breed differences measured in purebreed comparisons were very consistent with those measured on beef and dairy crossbreds. The Continental breeds and their dairy crosses gave rise to higher incidences of calving difficulty and calf mortality. They grew faster, converted their feed more efficiently and had carcasses with better conformation and less fat than the traditional beef breeds, their crosses, and the Friesian breed. The traditional Irish Hereford and Angus crosses were slower growing than the Friesian; they had better conformed carcasses but the Hereford crosses had fatter carcasses. Canadian Hereford and Angus bulls sired cattle which grew substantially faster than those sired by Irish bulls, but the Canadian Hereford carcasses had more fat.

1. Origin and use of beef breeds in Ireland

The cattle and beef industry makes an important contribution to the Irish economy in terms of home production, exports and also in employment, both within and outside of agriculture. In 1980 the cattle and beef industry accounted for about 36% of the total value of agriculural output, 50% of the total value of agricultural exports, and 19% of the total value of all exports. Approximately 85% of the output from the Irish beef industry is exported. Up to the 1970's the United Kingdom was the main market. This market demanded carcasses which were not too heavy and which had a fairly liberal fat cover. Hereford and Angus crosses were admirably suited for meeting these requirements. With Ireland's entry into the European Economic Community in 1973, new markets demanding heavier carcasses with more lean meat opened up. The Charolais, Simmental, Limousin, Blonde d'Aquitaine and Belgian Blue breeds were imported into Ireland to meet the requirements of these new markets. Extensive information on the origin of these breeds, their physical and functional characteristics and their breed organisation is given by French *et al.* 1966.

The Charolais was the first Continental breed to be imported The first animals arrived in 1964, and with further importations and the buildup of breeding stock, the nucleus pedigree breeding population now numbers about 750 breeding females. The Simmental were mainly imported from Austria, but smaller numbers were also imported from West Germany, Switzerland and France. Although this breed is largely bred for the dual purpose characteristics of milk and beef on the Continent of Europe, the animals imported into Ireland were intensively selected for high growth rate and good muscling to fulfil their role as a terminal crossing breed. The Simmental pedigree breeding population now numbers about 800 breeding females. The Limousin and Blonde d'Aquitaine breeds were imported around the mid 1970's. There are about 250 pedigree breeding females in each breed. The Belgian Blue is the latest Continental breed to be imported into Ireland. Twenty-four (6 bulls and 18 females) were imported in 1980. Irish Angus and Hereford were largely of British origin up to the late 1960's as frequent importations from Great Britain took place. Since then, however, in an effort to increase size, a policy of importations from Canada, combined with intensive selection within the breeds has been pursued. The estimated use of the different sire breeds in Ireland in the last decade is given in Table 1. These figures are based on artificial insemination figures and the numbers of bulls licensed for natural service. About 50% of cows are bred by AI and the remainder by natural service.

Beef bulls and beef inseminations accounted for 34% to 50% of matings, depending on year. There are two factors which largely explain the fluctuations in beef breed usage. Firstly, suckler cows, which would usually be mated to beef bulls, have declined in number from a high of 34% of the total cow population in 1974 to 22% in 1980, and secondly, the percentage of dairy

Table 1 Percentage of matings by sire breeds from 1970 to 1980

Year	Friesian and Shorthorn	Hereford and Angus	Continental	Others
1970	49.6	47.1	2.3	0.9
1971	48.6	48.0	2.5	0.8
1972	60.9	33.6	4.7	0.7
1973	66.6	25.3	7.5	0.7
1974	63.6	30.4	5.3	0.6
1975	54.9	38.9	5.5	0.7
1976	56.2	35.0	8.3	0.5
1977	57.6	32.7	9.3	0.3
1978	62.6	26.5	10.5	0.4
1979	64.7	23.8	11.2	0.2
1980	58.1	28.1	13.5	0.3

cows bred to beef bulls varies significantly between years.

1. Performance of pure breeds

Much of the information available on the pure beef breeds in Ireland comes as a by-product of the various recording and testing schemes making up the National Programme for the Genetic Improvement of Beef Cattle.

.1 On-farm weight records of pure breeds

Since 1975, weight data on about 15,000 purebred animals in about 460 herds have been recorded under the On-Farm Recording Scheme for pedigree cattle. Animals in each herd were weighed about every 100 days and weights were adjusted to the nearest 100th day of age on the basis of growth rate since the previous weighing. The data were analysed by the method of least squares and the effects of breed, sex, age of dam, month of calving, and some important two-way interactions were included in the model. The average age adjusted weights of the breeds, expressed as a percentage of the Hereford breed average, are given in Table 2.

The approximate numbers of animals recorded for the different breeds were : 8,000 Hereford, 1,100 Angus, 1,800 Charolais, 3,000 Simmental, 600 Limousin, and 500 Blonde d'Aquitaine. Weights at all ages were not recorded for all animals. This is especially true in the case of the older weights. In the case of the Limousin and Blonde d'Aquitaine breeds, only about 100 animals of each sex have been weighed at 300 days and 400 days. The figures cannot be taken as precise measures of genetic differences because of variations in the management and feeding systems under which the different breeds were reared. This can

Table 2 Age adjusted weights of pure beef breeds relative to the Hereford breed

Breed	Sex	Birth weight	100-day weight	200-day weight	300-day weight	400-day weight
Hereford	M	37.9	131	232	335	423
Average (kg)	F	34.3	116	184	240	284
Angus	M	90	97	94	92	92
	F	88	98	102	97	98
Charolais	M	118	116	117	116	116
	F	122	126	136	141	143
Simmental	M	111	122	121	115	117
	F	112	127	136	140	139
Limousin	M	100	105	104	100	101
	F	103	114	124	124	125
Blonde	M	115	111	108	108	115
d'Aquitaine	F	118	119	128	134	139

be seen clearly from the relatively higher performance of Continental breed females, due to the much higher planes of nutrition that they are generally reared on. However, the figures give a good guide as to the relative performance of the Continental breeds vis a vis each other, and also if the relative performance of males only are compared, an indication of how the traditional beef breeds compare with the Continental breeds can be obtained. The Charolais, Simmental and Blonde d'Aquitaine are substantially heavier than the Limousin. The Simmental tend to have lower birth weights than the Charolais and Blonde d'Aquitaine, but they catch up and are heavier by 100 days of age, probably because of the better milk supply from Simmental mothers. The Limousin breed have birth weights slightly heavier than Hereford and the weights at subsequent ages intermediate between Hereford and the other Continental breeds.

.2 Comparison of performance of pure breeds in Central Performance Test Station

More precise information on the relative performance of the different breeds is provided by data from the Central Performance Test Station. In Ireland, there is one performance test station where about 95 beef breed bulls are tested annually. There, selected bulls of the different breeds are evaluated under uniform conditions of feeding, housing and management. The performance of different breeds, expressed as a percentage of the Hereford breed average in the period 1976 to 1980, is presented in Table 3.

Table 3 Performance of pure breeds relative to the Hereford breed average in performance test station

Breed	No. bulls	Pre-test ADG (kg)	On-test ADG (kg)	Adjusted final wt (kg)	Feed conversion*	Withers height (cm)
Hereford average	177	1.18	1.17	517	6.43	112
Charolais	80	110	119	114	94	107
Simmental	135	113	121	116	98	109
Limousin	25	100	112	104	88	106
Blonde d'Aquitaine	22	103	122	113	85	110

* concentrate feed per kg liveweight gain

The bulls entered the testing station at 6 to 7 months of age and they were evaluated up to about 400 days. A feed consisting of rolled barley and unmilled dried grass suitably supplemented with minerals and vitamins was fed. The composition of the ration was about 84% DMD, 75% DOMD, 16% CP, and 9% ash. During the first three years bulls were fed to appetite three times a day and in the latter two years feed was available ad libitum. A small quantity of hay (about 2 kg/bull/day) was fed to help rumination and reduce digestive upsets. This was not taken into account when feed conversion figures were calculated.

Differences between the Hereford and the other breeds in pre-test average daily gain are somewhat lower than might be expected on the basis of on-farm weight records. This is probably explained by more intensive preselection of Herefords for central testing because of the larger numbers available. The Charolais, Simmental and Blonde d'Aquitaine were very similar for growth rate on test and adjusted final weight, and they were significantly superior to Limousin and Hereford. All the Continental breeds, particularly the Limousin and Blonde d'Aquitaine, converted their feed more efficiently than the Hereford. However, this must be interpreted in the context of the feed and feeding system the bulls were on, which was probably more suited to the bigger Continental breeds with their larger appetites and higher growth potential. Also, feed conversion was measured over a constant time period (200-400 days approximately) during the latter part of which the Herefords would have been laying down significantly more fat due to their earlier maturing nature.

3 Comparison of the carcass merits of pure breeds

A limited amount of carcass data was obtained on pure breed beef bulls which had to be slaughtered in 1980 due to a disease outbreak in the Central Perforamnce Testing Station (Table 4).

The bulls were fed and managed as outlined previously. The average age at slaughter was 500 days. The Charolais and Simmental breeds were very similar for all traits and they were significantly superior to the Hereford breed. The Limousin

Table 4 Comparisons of the carcass merits of pure breeds relative to Hereford

Breed	No. bulls	Carcass wt per day of age (kg)	Killout %	Conformation score*	Fat score**
Hereford average	24	0.728	58.2	5.7	4.6
Charolais	14	120	103	112	63
Simmental	27	119	102	111	67
Limousin	12	116	109	118	57
Blonde d'Aquitaine	7	122	110	118	37

* Visually assessed by experienced staff on the scale 7 (very good) to 1 (poor)

** Visually assessed by experienced staff on the scale 1 (leanest) to 7 (fattest)

had carcass gains per day of age approaching those of the Charolais and Simmental, largely because of their high killout percentage. The Blonde d'Aquitaine again had high carcass gains per day of age because of their high killout percentage and they also had extremely low fat levels.

. Performance of crossbreds from the dairy herd

More O'Ferrall and Cunningham (1977) summarised the relative performance of different breed crosses in Ireland from husbandry experiments and the beef progeny testing programmes carried out in the 1960's and early 1970's. In this section I will review results from the progeny testing programme since 1975.

.1 Ease of calving, calf mortality and gestation interval results

Ease of calving and level of calf mortality are important elements in the economic value of a particular cross. Extensive data on these characteristics have been collected by AI organisations in Ireland. These data provide a basis for looking at the relative merits of the different breeds as sires and dams.

Over 72,000 calvings from cows calving for the second or subsequent time have been surveyed since 1976. Calving data were recorded mainly by AI station personnel. Each calving was classified as 'normal', 'some assistance', or 'seriously difficult'. The general guideline was "if the farmer considers a calving difficult, then designate it 'seriously difficult' ". Data on calf mortality within 24 hours of birth were recorded. Calvings where presentation was abnormal, abortions, and multiple births were excluded from the analysis. Results for the different sire breeds are given in Table 5.

Table 5 Comparisons of sire breeds for ease of calving, calf mortality and gestation interval

Breed of sire	No. calvings	% some assistance	% seriously difficult	% mortality within 24 h	Gestation interval
Hereford	8,731	10.8	2.0	1.6	281.4
Angus	1,363	11.4	2.3	1.6	279.6
Charolais	7,286	16.7	5.0	2.9	283.9
Simmental	4,024	19.2	5.5	2.4	283.3
Limousin	862	14.3	6.1	3.1	283.3
Blonde d'Aquitaine	606	19.8	8.9	2.0	286.5
Friesian	49,416	10.2	1.7	1.8	280.7

In general, Continental breed sires gave rise to substantially more calving problems and calf mortality and longer gestation intervals than Hereford, Angus or Friesian. The levels of seriously difficult calvings for the Limousin and Blonde d'Aquitaine breeds are high relative to results from other countries. They are relatively new breeds in Ireland and they have been promoted as easy calving breeds, which probably led to them being used less selectively than the other fast growing Continental breeds. Selective mating, management and feeding of cows bred to Charolais and to a lesser extent Simmental bulls probably gave lower levels of calving difficulty and calf mortality for these breeds than if they were randomly mated. The widespread use of Angus sires on cows known to be problem calvers probably explains the relatively high incidence of seriously difficult calvings for this breed.

The calving results for the different dam breeds and crosses are given in Table 6.

Table 6 Comparisons of dam breed/crosses for ease of calving, calf mortality and gestation interval

Breed of cow	No. calvings	% some assistance	% seriously difficult	% mortality within 24 h	Gestation interval
Hereford X	2,701	12.3	4.4	1.9	282.4
Angus X	1,234	11.3	3.3	2.4	282.0
Continental X	422	14.7	5.0	2.3	283.1
Friesian	66,027	17.0	5.0	2.2	282.2
Shorthorn	2,375	14.2	3.9	1.7	282.5

No very significant differences were observed between the different breeds and breed crosses. Friesian and Continental cross cows tended to have a higher percentage of seriously difficult calvings. However, since different dam breeds/crosses generally exist in geographically separate areas where feeding and management may vary, there may be some confounding of management and feeding effects with breed or dam effects in these figures.

3.2 Comparison of Friesian and their beef breed crosses for growth rate and carcass merit

Information on the relative merits of Friesian and their beef breed crosses is available from the beef progeny testing programme for AI bulls. This programme is carried out by AI organisations in collaboration with the Department of Agriculture. Progeny groups by beef breed bulls were reared from birth to slaughter in a common environment with Friesian contemporaries at a number of test centres. The system of feeding, housing and management of the cattle was normal commercial practice. Spring-born calves were reared on milk replacers and meals. They were generally put out to grass around the end of May and limited quantities of concentrate were fed. Animals were housed during winter and generally fed on ad lib. grass silage with a small amount of concentrates. Age at slaughter varied from two years to two years and nine months, between test centres. Cattle at each test centre were slaughtered in one or two batches when it was considered that adequate numbers were finished. Approximately equal proportions of the progeny of each bull were slaughtered in each batch. The performance of crossbreds by the different beef breeds from the dairy herd relative to the average for Friesian contemporaries is summarised in Table 7.

Continental breed crosses had higher growth rates, higher killout percentages and better conformed carcasses with less fat than Friesian or Hereford and Angus crosses. These results are quite consistent with the results obtained for the pure breeds. As was the case with the Blonde d'Aquitaine pure breed, the crossbred progeny by the one bull which was tested had high killout percentages and low fat scores. Charolais crosses grew slightly faster than Simmental and their carcasses had better conformation and less fat.

Irish Hereford crosses had growth rates 2% less than Friesian. However, Canadian Hereford crosses grew 1% faster than Friesian. All Hereford crosses had higher conformation and fat scores than the Friesian and the Canadian Hereford crosses were substantially fatter than Irish Hereford crosses.

Irish Angus crosses grew substantially slower than the Hereford crosses or Friesian. However, the progeny of the one Canadian Angus bull which was progeny tested had liveweight gains equal to Friesian, but they had lower carcass gains per day of age due to a lower killout percentage. There were no differences between the Canadian and Irish Angus crosses for conformation score or fat score. The Angus crosses had slightly lower carcass conformation scores than the Hereford crosses but significantly lower fat scores. In fact, their fat scores were

Table 7 Growth rate and carcass characteristics of beef breed/Friesian crosses expressed as a percentage of the Friesian average

Breed of sire	No. sires	No. progeny	Liveweight gain per day of age	Carcass gain per day of age	Carcass length	Killout %	Conformation score*	Fat score**
Friesian average	105	1939	0.638 kg	0.345 kg	131.1 cm	54.09	4.14	3.07
Irish Hereford	25	383	98	98	98	100	111	115
Canadian Hereford	6	112	101	101	99	100	110	124
Irish Angus	5	77	94	92	97	99	108	99
Canadian Angus	1	17	100	97	99	97	108	99
Charolais	4	60	104	106	100	103	125	85
Simmental	14	252	103	104	100	102	112	96
Blonde d'Aquitaine	1	18	105	109	102	104	114	72

* Visually assessed by experienced staff on the scale 7 (very good) to 1 (poor)

** Visually assessed by experienced staff on the scale 1 (leanest) to 7 (fattest) slightly lower than those for Friesian.

References

French, M.M., Johansson, I., Joshi, N.R. and McLaughlin, E.A. 1966. European Breeds of Cattle 1 & 2. FAO agricultural study, No. 67, Rome. Food & Agriculture Organisation of the United Nations.

More O'Ferrall, G.J. and Cunningham, E.P. 1977. Review of Irish experiments on beef breeds and crosses. Proc. Seminar on crossbreeding experiments and strategy of beef utilisation to increase beef production. (Editors, I.L. Mason & W.Pabst). CEC, Luxembourg.

DISCUSSION ON DR. T. TEEHAN'S PAPER

Chairman: Dr. R. Thiessen

Òldenbroek: What was the distribution of age of cow in the data set on calving difficulties ?

Teehan: The data excluded first-calving cows.

Southgate: Do you have sufficient variation in fatness within breeds in the data used for Table 7 to quantify the relationship between fatness and other characters. For example, UK data has shown a regression of kill out percentage on subcutaneous fat percentage of 0.15. No doubt an alternative interpretation is possible. However, the data as presented yield the best commercial interpretation as they reflect the time of slaughter relative to seasonal and market variations.

Flanagan: Is the result in relation to crossbreds and purebreds in broad agreement ?

Yes, broadly speaking, the two are in agreement, with the possible exception of the Simmental carcass data. Also, ease of calving reflects direct and maternal effects and so some interaction may be possible there.

SIRE BREED INFLUENCE OF VARIOUS BEEF BREEDS ON CALVING PERFORMANCE, GROWTH RATE, FEED EFFICIENCY, CARCASS AND MEAT QUALITY

T. Liboriussen

National Institute of Animal Science, Rolighedsvej 25, 1958 Copenhagen V, Denmark

Summary

The improvements that can be obtained in regard to efficiency and quality by crossbreeding dairy cows with bulls from beef or dual purpose breeds are discussed.

The discussion is based on results from a beef x dairy crossbreeding project in which 15 European beef or dual purpose breeds were tested as sire breeds in crossbreeding with Danish Red, Danish Friesian, and, to a lesser extent, Jersey cows.

When crossbreds were tested in intensive young bull production, the highest growth rates were obtained by Charolais crossbreds, which had approximately 9% higher growth rate than young bulls of "dairy" type. Highest carcass quality was obtained from crossbreds after breeds represented by double-muscled bulls (Piedmont, Belgian Blue-White, and Blonde d'Aquitaine), but also Limousin crossbreds had very high carcass quality.

The high birth weight of crossbreds from large framed beef breeds with very high growth capacity causes calving difficulties, if these breeds are used for crossbreeding with cows of large dairy breeds. The best suited beef breeds for crossbreeding to large dairy breeds are consequently Limousin or Piedmont.

Introduction

The aim of this contribution is to quantify the improvements in efficiency and quality of beef production that can be obtained by crossbreeding cows of large dairy breeds to bulls of various beef and dual purpose breeds. The influence of sire breed on calving performance of the dairy breeds will also be considered.

Material

The material background is a beef x dairy crossbreeding project in which 15 European beef and dual purpose breeds were tested in crossbreeding with Danish Friesian and Danish Red. The project was presented by Andersen and Liboriussen (1977), so only the main features are given here.

The breeds were tested in two series, with eight breeds in each.

The first series included Charolais, Simmental, Danish Red

and White, Romagnola, Chianina, Hereford, Blonde d'Aquitaine and Limousin. The calves were born in autumn 1972, 1973 and 1974.

Calves for the second series were born in the autumn of 1975, 1976 and 1977. The sires of these calves were AI bulls from the following breeds : Charolais, Aberdeen Angus, Gelbvieh, Piedmont, South Devon, Braunvieh, West Flemish Red and Belgian Blue-White.

Each sire breed was represented by 3-5 AI bulls.

The calves were born on private farms in Jutland. Calving difficulty and stillbirth were recorded by the farmers by means of questionnaires. The method and some preliminary results were presented by Liboriussen (1977).

At the age of 2-4 weeks the calves were gathered at experimental stations where they were tested in various production systems, as described by Andersen and Liboriussen (1977).

Crossbreds from both series were tested in intensive bull production at the central breeding station Egtved. The animals were tied up, and their individual feed intake was recorded. Feeding was based on concentrates.

In Series I all animals were fed according to appetite. In Series II each breed group was distributed on 3 feeding levels (ad lib. 85% and 70%), and those that were fed restrictively were further distributed on two treatments with different protein levels.

Serial slaughtering was used in both series. In Series I the animals were slaughtered at eithter 320 kg, 440 kg or 560 kg.

The carcass and meat quality were examined by the Danish Meat Research Institute, and included :

(a) Standardised slaughtering with weighing of carcass, organs, and scoring of the carcass.
(b) Commercial cutting, followed by tissue separation of the right side of each carcass.
(c) Physical and chemical measurement and taste panel evaluation of m. longissimus dorsi.

The results from young bull production of the crossbreds from Series I can be compared with results of purebred Danish Red and

Danish Friesian from the ordinary progeny tests, since the feeding systems were identical. In Series II, purebred Danish Red and Danish Friesians were tested together with the crossbreds.

Results

The influence of sire breed on calving performance and birth weight is summarised in Table 1 (Series I) and in Table 2 (Series II). A direct comparison of sire breed results obtained in different series is not possible. The reason for this is that Jersey was included as the dam breed in Series I and not in Series II. As can be seen from Table 1, the Jersey cow generally has a greater maternal influence on calving performance than cows of Danish Red and Danish Friesian

Table 1 Influence of sire breed on frequency of difficult calvings, perinatal mortality and birth weight Series I

Sire breed	No.	Difficult[1] calvings %	Perinatal[2] mortality %	Birth weight kg
Hereford	134	31	6.0	36.3
Danish Red-White	140	38	5.0	40.1
Limousin	113	39	6.2	38.7
Simmental	141	41	0.7	41.6
Bl. d'Aquitaine	101	43	5.9	41.1
Chianina	127	53	6.3	42.9
Charolais	114	62	8.8	45.1
Romagnola	134	72	7.5	43.6
Dam breed				
Danish Red	454	50	5.5	46.3
Danish Friesian	420	52	6.4	44.6
Jersey	130	21	5.4	32.6
Total	1004	47.3	5.9	41.1

[1] Pulling assistance or veterinarian assistance

[2] Including calves which died within 48 h after birth.

Normally, 30-35% of cow-calvings with Danish Red or Danish Friesian are assisted.

In Series I all sire breeds, except Hereford, had a negative influence on calving performance. This negative influence is mainly caused by an increase in birth weight. Crossbreds with Chianina, Romagnola and Charolais had average birth weights, which were more than one standard deviation higher than the average birth weight for purebred Danish Red, Danish Friesian and Jersey. Serious calving problems can be expected if these breeds are used for crossbreeding with large dairy herds, even if - as in this case - they are used on cows.

The differences in perinatal mortality were not significant, but this indicated mainly the watchfulness of the farmers.

From the breeds tested in Series II a strong negative influence on calving performance was observed for Belgian Blue-White and Charolais. Aberdeen Angus was the only breed in this series with fewer calving problems than are usually found in purebred Danish Red and Danish Friesian.

Table 2 Influence of sire breed on frequency of difficult calvings, perinatal mortality and birth weight Series II

Sire breed	No.	Difficult[1] calvings %	Perinatal[2] mortality %	Birth weight kg
Aberdeen Angus	77	13	2.6	34.1
Braunvieh	86	36	3.5	42.9
Gelbvieh	60	40	6.7	41.5
South Devon	76	49	5.3	40.2
West Flemish Red	74	53	10.8	48.6
Piedmont	82	55	9.8	42.2
Belgian Blue-White	69	61	8.7	44.8
Charolais	90	61	5.6	47.6
Dam breed				
Danish Red	279	44	6.8	41.9
Danish Friesian	335	48	6.3	42.0
Total	614	46.2	6.5	41.9

[1]Pulling assistance or veterinarian assistance.

[2]Including calves that died within 48 h after birth.

Results concerning growth rate, feed efficiency and carcass quality of young bulls are summarised in Table 3 (Series I) and Table 4 (Series II).

Table 3 Beef production from crossbred young bulls compared with the expected production of purebred Danish Red and Danish Friesian under similar conditions. (Weight at slaughter = 450 kg)

	Av. daily gain (g)	Sc. f.u./ kg gain	Dressing %	Confor- mation	Lean %
Charolais x	1338	3.57	54.4	8.0	71.2
Bl. d'Aquitaine x	1338	3.62	55.8	8.3	74.1
Simmental x	1310	3.73	53.9	7.6	70.4
Danish Red-White x	1304	3.75	53.9	7.8	69.6
Romagnola x	1302	3.81	54.1	7.7	70.3
Chianina x	1295	3.76	55.0	6.1	70.9
Limousin x	1247	3.99	56.0	9.2	70.9
Hereford x	1223	4.02	54.2	7.6	65.9
RDM, SDM[1]	1225	4.05	53.7	7.0	69.0

[1] Average of Danish Red and Danish Friesian

The ranking of breeds according to growth rate and according to feed efficiency is almost identical, as a consequence of the strong genetic relation between these traits.

The Charolais crossbreds obtained the highest growth rate in both series. Their superiority over purebred Danish Red and Danish Friesian was of the order of 9%. Blonde d'Aquitaine crossbreds had also a very high growth rate in this production system, but not in extensive production systems. crossbreds with Simmental, Danish Red-White, Romagnola, Chianina, West Flemish Red, Belgian Blue-White, and South Devon had all between 3% and 7% higher growth rate than purebred Danish Red or Danish Friesian. Only the Angus crossbreds had substantially lower growth rate than purebred Danish Red and Danish Friesian, while Braunvieh, Gelbvieh, Piedmont, Hereford and Limousin crosses obtained results similar to Danish Red and Danish Friesian.

Table 4 Beef production from crossbred young bulls, compared with the production of purebred Danish Red and Danish Friesian.
Series II (Average weight at slaughter = 441 kg)

	Av. daily gain (g)	Sc. f.u./ kg gain	Dressing %	Confor-mation	Lean %
Charolais x	1323	3.97	55.0	8.7	72.4
West Flemish Red x	1269	4.13	54.6	7.9	69.2
Belgian Blue-White x	1262	4.19	56.7	8.9	73.3
South Devon x	1255	4.11	53.4	7.3	68.4
Piedmont x	1217	4.19	57.6	8.6	75.1
Gelbvieh x	1210	4.30	54.3	7.7	69.7
Braunvieh x	1173	4.38	54.8	8.5	68.4
Aberdeen Angus x	1103	4.83	54.5	8.5	65.5
RDM, SDM[1]	1213	4.35	53.2	6.6	67.9

[1] Average of Danish Red and Danish Friesian

Carcass quality: Three of the sire breeds were represented by double-muscled bulls, viz. Piedmont, Belgian Blue-White and Blonde d'Aquitaine, and crossbreds after these breeds had very high dressing percentage, and their carcasses had a very muscular appearance (good conformation scores). Furthermore, they had a very high lean-to-bone ratio. The Limousin crossbreds also had a very high dressing percentage and lean-bone ratio, and they also obtained very high gradings for conformation. They did, however, deviate from the crossbreds after the double-muscled breeds in regard to lean content of the carcass. Obviously the Limousin crossbreds were earlier maturing than the others.

Crossbreds after the large framed late maturing beef breeds - Charolais, Chianina and Romagnola - had higher lean percentage higher dressing percentage and - except Chianina - they also obtained higher conformation scores than purebred Danish Red and Danish Friesian. Their superiority in these respects was most pronounced when the slaughter weight was high (Liboriussen, 1978).

Crossbreds after the dual-purpose breeds - Simmental, Danish Red-White, West Flemish Red, Gelbvieh and Braunvieh - had only slightly better carcass quality than Danish Red and Danish Friesian. For most of the carcass quality traits the observed difference is less than one genetic standard deviation.

The early maturity of Hereford and Angus is reflected in low lean percentage of crossbreds out of these breeds. That - combined with low growth rate - makes these types less fitted for production of young bulls under Danish conditions.

Meat Quality: The observed differences in meat quality traits have generally been small, and the meat quality can be ignored in a discussion of beef versus dairy breeds.

Discussion

The suitability of a breed for commercial crossbreeding for beef production with dairy breeds is mainly determined by its influence on calving performance of the dairy cows, and the superiority of the crossbreds, relative to the "dairy" type for beef production traits.

I have previously calculated the economic importance of the observed differences in calving performance, growth rate, feed efficiency and carcass quality between the various crossbreds from Series I, and purebred Danish Red and Danish Friesian. The results are summarised in Table 5.

In these calculations the risks of calving difficulties were considered only if they gave reason for increased veterinarian assistance of calving, or increased perinatal mortality. The increased management problems and the negative effect on cow productivity that is also connected with calving difficulty, have not been considered. Not because they are unimportant, but simply because it is very difficult to quantify their economic importance. In reading Table 5, one has therefore to keep in mind that the reproduction costs of crossbreds from the large beef breeds are underestimated. That is particularly the case for the Blonde d'Aquitaine crossbreds.

Table 5 The influence of genotype on net profit from production of young bulls, slaughtered at 450 kg liveweight. D.kr. (Liboriussen 1978)

	Reproduction costs[1]	Production costs[2]	Sales price[3]	Net profit	
				Per head	Per year
Bl. d'Aquitaine x	596	2138	3604	870	977
Limousin X	619	2350	3674	705	736
Charolais x	711	2107	3500	682	774
Simmental x	636	2194	3452	622	686
Danish Red-White x	646	2206	3454	602	662
Chianina x	664	2195	3448	589	648
Romagnola x	700	2238	3467	529	580
Hereford x	615	2386	3468	467	477
RDM, SDM	610	2374	3390	406	417

[1] Reproduction costs = $\frac{1}{a_j} \cdot (550 + b_j \cdot 450)$, where

a_j = proportion of born calves, that were alive 16 days after birth, of the jth genotype

b_j = proportion of calvings with veterinarian assistance

[2] Production costs = feed costs + fixed costs per day + interest costs

[3] According to carcass weight and grading.

The production costs estimate the economic value of growth rate and feed efficiency. With Charolais crossbreds, which were the most superior in these respects, the production costs involved in the production of 450 kg young bulls is approximately 13% lower than with purebred Danish Reds or Danish Friesians.

The sales prices were calculated on the basis of carcass weight and conformation with the price differentiation that is generally used for young bulls in Denmark. Due to their superior conformation and high dressing percentage, the Limousin crossbreeds obtained the highest price per kg liveweight. Their superiority over purebred Danish Red and Danish Friesian was of the order of 8%.

The highest increase in net profit per young bull was obtained by Blonde d'Aquitaine, followed by Limousin and Charolais. Due to underestimation of the costs of calving difficulties Limousin is probably to be preferred for cross-breeding with Danish Red and Danish Friesian. The increase in profit corresponds to approximately 300 D.kr. per young bull.

Similar calculations have not been made for the crossbreeds from Series II. Of the breeds tested in this series, the Piedmont breed is probably the one that comes nearest to Limousin.

The genetic capacity for growth in the large dairy breeds, like the various strains of Friesians and the Red Scandinavian dairy breeds, is not much different from that of many specialised beef breeds. Even by crossbreeding with the breed having highest growth capacity - the Charolais - only approximately 10% increase in growth rate can be expected.

The strong genetic relation between pre- and post-natal growth capacity implies that breeds with the highest growth capacity also give rise to most calving difficulties, due to high birth weight. Crossbreeding cows of large dairy breeds with bulls of the breeds with highest growth capacity, is consequently not advisable, unless they give relatively small calves.

The genetic distance between the large dairy breeds and some of the specialised beef breeds is greater when we look on carcass quality traits. Crossbreeding with double-muscled bulls gives crossbreds with a dressing percentage of 8% higher than that of purebred "dairy" animals, and the genetic coefficient of variation within breeds is of the order of 1.5%.

The increase in carcass value that can be obtained is, however, relatively small, and in my opinion not high enough to encourage dairy farmers with cows of the large dairy types to use commercial crossbreeding.

The situation is quite different if we take a small dairy breed like the Jersey. The growth capacity of this breed is less, and the increase that can be obtained by crossbreeding consequently higher. Cows of this breed can also give birth to relatively large calves, without serious calving difficulties.

Low value of culled dairy cows will tend to decrease the optimal culling rate. The surplus of reproductive capacity is consequently higher in a small dairy breed than in a large dairy breed. That will also make it more worthwhile to use commercial crossbreeding in Jersey herds.

Conclusion

The best compromise for crossbreeding with large dairy breeds seems to be breeds with very high carcass quality and moderate growth capacity such as Limousin or Piedmont.

The increase in profitability of beef production that can be obtained, seems, however, not to be high enough to encourage farmers with large dairy cows to make use of commercial cross-breeding for beef production.

Dairy farmers with small dairy cows - like Jersey - are in a different situation, because :-

1) The surplus of reproductive capacity in such herds is generally higher.
2) The superiority of the beef x dairy genotype is higher.
3) The risk of calving difficulty is lower.

References

Andersen, B. Bech and Liboriussen, T. 1977. "Danish cross-breeding experiments with beef breed on dairy and dual-purpose breeds." Proceedings of EEC Seminar on Cross-breeding experiments and strategy of beef utilisation to increase beef production, Verden, Federal Republic of Germany, 1976. EUR 5492 e.p. 230-239.

Liboriussen, T. 1979. "Influence of sirebreed on calving performance, perinatal mortality and gestation length." In "Current Topics in Veterinary Medicine and Animal Science." Vol. 4, p. 120-131.

Liboriussen, T. 1978. "Brugskrydsning i SDM og RDM". 466. Beret. fra Statens Husdyrbrugsforsøg, København, 123 pp.

DISCUSSION ON DR. T. LIBORIUSSEN'S PAPER

Chairman: Dr. R. Thiessen

Jansen: Did farmers know the identity of the breed when they were scoring calving assistance ?

Liboriussen: No, it was coded, but there is the possibility that they learned the code after a while. There is also little doubt that many of the cows that were offered assistance did not need it.

Fewson: What was the distribution of the age of cows for the calving difficulty data ?

Liboriussen: All cows were second calvers or older, but the exact age was not known.

Foulley: What was the sampling procedure used for sires, and were all bulls progeny tested ?

Liboriussen: Mostly five sires were used per breed, but only three were used for the Blonde d'Aquitaine, due to unavailability. All were AI bulls from which semen was available. All sires were sampled on the nucleus in the country of origin. Herefords and Charolais were from the Danish Hereford and Charolais populations, however. They were not all progeny tested because progeny test results were not available for all populations. Results were used where available.

Boucque: How was dressing percentage calculated ?

Liboriussen: This was based on liveweight in the morning, before feeding, and cold carcass weight after a short journey to the factory.

Oostendorp: The level of calf mortality in the Piedmont shown in Table 2 seems high. Our experience is that it is of the order of 2 - 3%.

Liboriussen: There is no plausible explanation, except the numbers were too small. The differences were not, in fact, significant.

Southgate: I seem to recall that Milk Marketing Board mortality data based on 200 calvings per breed, provided a least significant difference of 3%, so the differences you would expect to show as significant would be larger.

COMPARISON OF DIFFERENT SIRE BREEDS CROSSED WITH FRIESIAN COWS : PRELIMINARY RESULTS

F. Menissier, J. Sapa, J.L. Foulley, J. Frebling and B. Bonaiti

Station de Génétique Quantitative et Appliquée,
Centre National de Recherches Zootechniques,
I.N.R.A. 78350 Jouy-en-Josas, France.

Summary

Considering the present degree of dairy specialisation in cattle herds in Europe, terminal crossing appears as one of the most efficient means of improving their overall productivity (milk plus beef). Therefore, under the financial support of the EEC, 17 breeds or strains were tested in France in order to compare their sire value in crossing with Friesian cows. Data from this experiment, involving respectively 1,402 calves and 428 bulls, were analysed for birth characteristics of calves on the farm, and for intensive fattening and slaughter performance of bulls at the station.

Although influenced by within-breed sampling of sires, the results indicate that : (1) Terminal crossing significantly improves the efficiency of fattening and carcass performance, but increases simultaneously birth weight and therefore the frequency of dystocia. (2) Even though no breed was systematically better than the others, large differences were shown in growth potential, muscling and maturing pattern. (3) British breeds do not have as unfavourable an influence on the rate of dystocia but continental beef breeds and related specialised strains improve most the efficiency of fattening; dual purpose breeds are rather similar to the latter for growth potential but slightly lower as regards muscling.

These results should help in the choice of the most suited sire breeds for different production systems, so that an optimum balance between milk and meat production could be rapidly reached within the EEC.

Introduction

Industrial crossing of dairy cows with beef bulls is one of the most efficient and most rapid means to increase the productivity of dairy herds (Anderson & Lindhe, 1973 ; Cunningham, 1974; Cunningham & McClintock, 1974; Menissier et al, 1974; Elsen & Mocquot, 1976). This crossing allows an increase in their meat production, maintaining and even improving at the same time their milk production. The advantage is naturally connected with a higher carcass value of crossbred calves compared to those of the dam pure breed, whereas the unfavourable side or the limits of this crossing depends on its secondary effects on the cows' milk production and management (calving difficulties in particular) as well as on the crossing rate in these herds which depends on the replacement rate of females.

Industrial crossing has therefore been a widely used production system in France for more than 25 years in some of the dairy herds (Vissac et al., 1959, 1965, 1971; Poudardieu & Vissac, 1968; Foylley et al., 1975). It has mainly been developed in the Central and South Western part of France, on small sized farms, which have not specialised in milk production and generally producing veal calves. Owing to industrial crossing with French beef breeds, a part of these herds and of this female population composed of hardy, dual purpose and dairy breeds, as well as of crossbred females, has been reconverted into meat production. This is why industrial crossing experiments were set up in France since 1960 to compare out main beef breeds used in veal production, (Vissac et al., 1971; Foulley, et al., 1978) and later in the production of young bulls (Frebling et al., 1970). These comparisons showed the economic superiority of Charolais crossbred calves as compared to Blond d'Aquitaine and Limousin crossbreds, especially because of the interest that the farmers show in the carcass conformation of the calves and also because of local market practices each of these three sire breeds has been widely used in industrial crossing.

During the same period, the rapid development of artificial insemination in these small sized farms contributed to a large diffusion of industrial crossing with specialised beef breeds. Thus in 1970 and 1975, around 2,614,000 and 2,050,000 cows respectively were inseminated with semen from Charolais, Limousin and Blond d'Aquitaine bulls (Foulley & Menissier, 1981). Specific selection schemes for industrial crossing were therefore set up in France from 1959 to 1970 with the aim of obtaining a maximum genetic change in beef value of bulls intended for artifical insemination (Vissac, 1972; Frebling et al., 1972; Foulley, 1976; Foulley et al., 1978). These selection schemes have little by little been more and more integrated (Frebling et al., 1972) then optimised (Mocquot 1972; Mocquot & Foulley, 1973) and have led to the creation of specialised sire lines for terminal crossing (Vissac, 1972; Foulley, 1976; Bibe et al., 1977; Menissier, 1980). Their maximum efficiency was reached around 1970-74 (Gaillard et al., 1974).

At that time the overall economic crisis and especially that affecting beef production, a new context rapidly occurred which changed the condition of the development of industrial crossing. In particular the greater specialisation of dairy herds towards milk production was immediately accompanied, on the one hand, by a reduction in the role of industrial crossing in these herds and consequently in the number of inseminations from beef bulls and, on the other hand, by a very large development of the commercialisation of "non-finished" calves, either at birth ("8 day calves") or at weaning ("feeder calves"). These changes have deeply affected the context of the selection of A.I. bulls intended for industrial crossing (reduction of the extent of the selection schemes, modification of selection goals, changes in testing methods) (Foulley & Menissier, 1981).

Because of this new context for industrial crossing in French dairy herds and further development of specialisation of milk production within all dairy populations in Western Europe, it seemed to us to be necessary to estimate objectively the improvement obtained in these dairy populations by the industrial crossing. This concerns in particular the Friesian population where the specialisation and milk yield increment are accompanied by a lower meat productivity. The sire breeds liable to be evaluated for industrial crossing with Friesian dairy cows, come from several types of populations:

- The continental beef breeds (French and Italian) and their specialised types or lines for terminal crossing, because of their large muscle development and high growth potential;
- The large sized dual purpose breeds because of their good growth rate as well as those (Belgian and Italian) using the double muscle trait;

The British beef breeds which, contrary to the former, have a more favourable influence on calving difficulties.

This is the objective of the composition study concerning the European main sire breeds used in industrial crossing with Friesian cows, partly financed by the Standing Committee of Agricultural Research of the E.C.C. (Tayler, 1976) and performed from 1975 to 1979 in France. We shall here present some preliminary results.

Experimentation and Method of Analysis

In this experiment calves were obtained in dairy herds from Friesian cows crossed with 18 sire genetic types used in this comparison:

(1) Sire breeds and genetic types compared:

The number of sire breeds was restricted to 18 for economical reasons. The sire breeds were chosen among the the previously indicated different types of cattle populations. They may be grouped in several sets according to the trend of their selection:

(a) Breeds of terminal crossing type (or "T.C" type,) among which we considered 3 groups of different breeds or genetic types:

*French beef breeds, specifically selected for crossing (n=3) - Charolais (CH_T), Limousin (LI_T) and Blond d'Aquitaine (BA_T); only the animals selected for their carcass traits were considered.

*Synthetic sire lines specialised in terminal crossing and selected independently of their parental breeds (n=3) (Foulley, 1976; Menissier, 1980) i.e. "Coopelso 93 (C93) issued from CH_TxBA_T) and the two double muscled sire lines "Inra 95" issued from CH_TxBA_T ($I95_{CP}$) and MAxBA_TxLI_T ($I95_{MB}$).

*Dual purpose breeds (n=2) - the Belgian Blanc Bleu (BB) and the Italian Piemontese (PI) where the double muscle trait has been searched for and selected for improving their beef production ability.

(b) non specific breeds for terminal crossing or breeds for pure breeding type (or "P.B" type), whose selection is directed towards beef production in purebreeding systems, among these are three groups:

*Continental beef breeds selected for their meat production ability and their maternal abilities as well as reproductive traits; these breeds are : the Chianina breed (CI) from Italy and the three specialised French beef breeds - Charolais (CH_E), Limousin (LI_E) and Blond d'Aquitaine (BA_E) but for the latter only animals selected for utilisation in pure breeds.

*Large sized dual purpose breeds (n=2), Maine-Anjou) (MA) and French Simmental or Pie-Rouge de l'Est (SI), selected for beef and milk production and used in beef or dairy herds.

*British beef breeds selected for meat and maternal abilities; we used South Devon (SD) and Hereford from the United Kingdom (HE_{UK}) as well as Hereford from the United States imported by French Association of Hereford breeders (HE_{US}).

(c) the Friesian breed of the N.F.S. type (FR) was used as a reference sire breed.
A sample of 6 bulls was chosen from each of the sire breeds to produce the experimental calves by A.I. The bull sampling varied according to groups of breeds. As regards the specialised French beef breeds for T.C. (CH_T, BA_T, LI_T) and the specialised sire lines (C93, 195_{MB}) the bulls were chosen among those in the best position after performance testing, but not yet progeny tested and selected. On the other hand, in all the other breeds of the P.B. or T.C. types, the bulls were chosen among the best of those recommended for selection in their breed and which had successfully undergone all the steps of selection, in particular the progeny testing on beef

value in some bulls (PI, BB, CH_E, BA_E, LI_E, MA, SI, HE_{UK}).

(2) Management of experimentation:

Two annual series of A.I. (1976 and 1977) were organised with three bulls per sire breeds chosen at random in the sample of 6 bulls, for each series (54 bulls/year). From mid March to early June, 3,565 contract matings were made on Friesian cows in small private farms where industrial crossing was currently practised with French beef bulls. These farms are located in the A.I. areas of 2 semen production units (Union of A.I. Centres) : one for the Centre ("Union Auvergne-Limousin-Charente" or U.A.L.C.) and the other for the South-West (Union "Midatest") of France (Figure 1). Each bull (n=108) was used in these two areas. Most of the cows used in this industrial crossing experiment were adult (84% at third calving and more) and of the Friesian type without any Holstein" blood.

The calving conditions were evaluated by the farmers according to the national system of recording of these performance : the score of difficulty (1 = without assistance, 2 = light assistance, 3 = important assistance, 4 = caesarian section), the date and weight (estimated or determined by weighing on a scales, at birth of calves were given by the farmers together with their identification. To check the sample of cows used, their chest circumference was recorded by an operator when the calves were purchased. Birth conditions of 1,518 calves were recorded and 1,402 births were analysed after elimination of twin births and incomplete records.

The calves were purchased immediately after birth and then raised in nursery (controls started at an average age of 20 days) like traditional baby calves ("8 day calves"). All calves were purchased except those exhibiting sanitary problems, i.e. 795 males and 697 females, and placed in six nurseries distributed in the birth areas of the calves (Figure 1). They were managed in the classical way like

FIGURE 1. Location of insemination areas, nurseries, performance stations and slaughterhouse used in the EEC-INRA crossbreeding experiment.

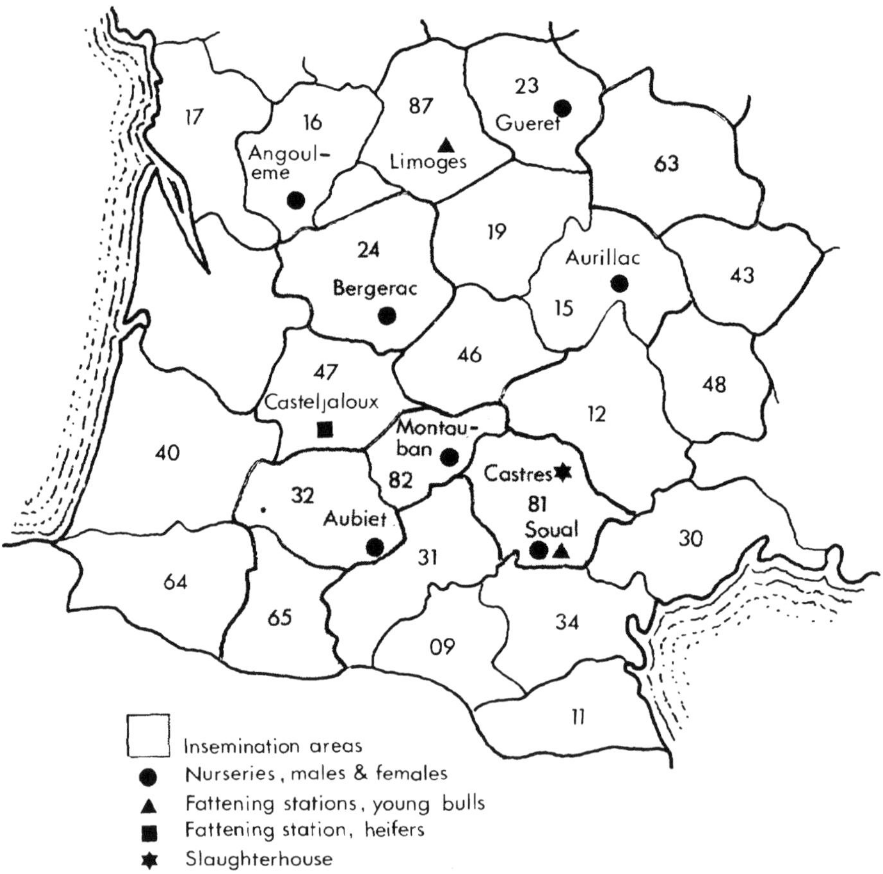

calves intended for breeding, with a restricted growth rate and weaned at the age of 4-5 months. Growth, morphology and conformation tests were performed in these nurseries. Towards the age of six months, 575 males were transferred to fattening stations while 359 females calves were either grouped in another fattening unit (feeding based on grass and maize silage) for the first set or sold to feeders because of the high experimental costs for the second set.

439 males calves were intensvely fattened after weaning and controlled in two performance testing stations : one in Limoges (120 tied up) and the other at Soual (50 tied up and 50 per box of five animals). The male calves were distributed randomly in these stations according to their sire and area of origin. They were fed dried lucerne and barley (70 and 30p 100, respectively). About 2/3 of these young bulls chosen at random from among each sire progeny were fed ad libitum already from the age of 8 months; after a given fattening period, half of the animals taken at random were slaughtered at the age of 14 months and the other half at 17 months. The remaining young bulls (1/3) received the same diet which was restricted and determined according to the metabolic weight whatever their genetic type. The purpose of this restriction was to reduce feed intake of the early maturing genetic types of bulls (FR, HF_{US}, HF_{UK}, SD) (10-15p 100) more than that of the bulls belonging to the other later maturing genetic types (less than 10p. 100) (CH_T, CH_E, BA_T, BA_E, LI_T, LI_E, C_{93}, 195_{CB}, 195_{MB}, BB, PI, MA, SI, CI). We attempted to slaughter these restricted bulls at a weight comparable to that of the <u>ad libitum</u> bulls at 14 months for the early maturing types and at 17 months for the later maturing types; finally, they were slaughtered 14 days later. The other male calves (n=136) which were not tested in these stations were either fattened in a fattening unit (maize or grass silage and slaughtering at constant weight) for the first set (n=136) or sold after weaning and not tested for the second set (n=96).

All the young bulls were slaughtered (at constant age for most of them) in an industrial slaughterhouse close to one of the stations (Soual) but 300 km from the other (Limoges). The bulls were slaughtered (n=432 + 133) as soon as they arrived in the slaughterhouse. The controls concerned the slaughter performance (carcass and live weight), carcass quality, (measures and conformation according to EAAP recommendation, De Boer et al., 1974), carcass composition estimated from the 6th rib cut, weight of internal fat and canon bone (Robelin & Geay, 1975) as well as some meat characteristics measured on the longissimus dorsi muscle (pH, water retention, shearing force and reflectance values after six days of storage, collagen and lipid contents, hoeminic iron after freezing). A sample of 70 half-carcasses were also subjected to a commercial cut.

(3) Analysis of results

The statistical analyses of these data concerned almost exclusively, on the one hand, birth conditions and performance in the nursery of male and female calves, and on the other hand, fattening and slaughter performance of the young bulls fattened in the performance test stations according to the *ad libitum* system. The analysis was made using the following mixed linear model :-

$$Y_{ijkl} = \mu + a_i + g_j + \rho_{ijk} + c_k + \sum_{\alpha} (\beta_{\alpha} \cdot x^{\alpha}_{i}jkl) + e_{ijkl}$$

Where μ = mean,

a_i = fixed effect of the ith year (i=1,2);

g_j = fixed effect of the jth genetic type (j=1,...,18);

p_{ijk} = random effect of the kth sire of the jth genetic type used the ith year (k=1,2,3);

c_k = all the other fixed effects crossed with a_i and g_j (region, nursery, station, parity, sex, sex x parity) etc...) variable according to the y performance analysed;

x^{α}_{ijkl} = αth covariable of the model (birth date, chest circumference of dam, age at the beginning of fattening);

e_{ijkl} = random errors, independent, of means equal to zero and of homogenous variances.

Estimation of fixed effects and estimable linear functions were obtained by solving the generalised least square equations (BLUE estimators) involving a value of the ratio σ_e^2/σ_ρ^2. This ratio being known, the corresponding hypothesis tests were then Student's t or F tests in our case. According to Henderson and Henderson (1979) the a priori unknown value of the ratio was chosen on the basis of estimations of the variance components σ_ρ^2 and σ_e^2 obtained either by Method 3 or by a procedure of calculation of the restricted maximum likelihood using the solutions of the generalised least squares (Schaeffer, 1979; & Thompson, 1979).

Birth Conditions of the Calves

In order to estimate the importance of calving difficulties we estimated the sire breed effects (Table 1) after correction of the other effects (year, region, birth date, chest circumference and parity of the dam, sex of the calf, sex x parity) on the score for birth difficulties and frequency of assisted births (scores 2, 3 and 4) and of difficult or dystocic births (scores 3 and 4). Whatever the criterion used the results agree and the following phenomena should be emphasised:

(1) It is clear that almost all sire breeds for crossing studied increase the birth difficulties. The score for birth difficulties increased by +0.19 points, the frequency of assisted births by +12.3 p.100 and that of dystocic births by +7.9 p.100. This increment of birth difficulties is associated with a higher birth weight of the calves (+4.3 kg) and a longer gestation length (+5.7 days). We were unable to separate the role played by heterosis effects from that played by additive genetic effects. As shown previously (Menissier & Belic, 1968; Philipsson, 1977; Menissier and Foulley, 1979; Foulley & Menissier, 1981), it is interesting to notice (Figure 2) the existence of a certain threshold effect of the birth weight of calves on birth difficulties, which

TABLE 1: Sire Breed Effect on Birth Conditions of Calves out Mature Friesian Cows – (figures are expressed in deviation to the Friesian effect).

SIRE BREEDS OR MAIN EFFECTS:	Number	DIFFICULTY SCORE (1 to 4) : points	P.100 OF CALVINGS : ASSISTED : (%)	P.100 OF CALVINGS : DYSTOCIC : (%)	BIRTH WEIGHT (KG)	GESTATION LENGTH : (DAYS)
CHAROLAISE	86	-0,30±0.12	+17,0±9.3	+11,9±5.2	+5,35±0.93	+5,3±1.2
BLONDE D'AQUITAINE	82	+0.13±0.12	+ 6,1±9.4	+ 6,5±5.3	+2,82±0.95	+7,3±1.2
LIMOUSINE	81	+0,17±0.13	+11,9±9.4	+ 5,3±5.3	+4,04±0.96	+7,8±1.2
"COPELSO 93"	80	+0,21±0.13	+15,1±9.4	+ 6,4±5.3	+5,85±0.96	+7,5±1.2
"INRA 95" (MA,CH)	73	+0,27±0.13	+15,9±0.5	+10,0±5.4	+5,70±0.98	+6,0±1.2
"INRA 95" (MA,BA,LI)	90	+0,30±0.12	+14,8±9.3	+14,3±5.2	+4,21±0.94	+5,7±1.2
BLANC BLEU BELGE	67	+0,25±0.13	+17,8±9.6	+ 5,4±5.4	+4,82±1.02	+2,3±1.2
PIEMONTESE	67	+0,27±0.13	+16,4±9.6	+10,9±5.4	+4,31±1.00	+7,4±1.2
CHAROLAISE...............	66	+0,32±	+21,7±9.8	+11,3+5.5	+6,37±1.00	+5,1±1.3
BLONDE D'AQUITAINE	86	+0,09±	+ 3,5±9.4	+ 6,0±5.3	+3,17±0.96	+7,3±1.2
LIMOUSINE	81	+0,03±	+ 0,8±9.4	+ 2,0±5.3	+3,09±0.97	+7,6±1.2
CHIANINA	86	+0,36±	+20,0±9.4	+14,2±5.3	+5,56±0.96	+7,4±1.2
MAINE-ANJOU	75	+0,30±	+18,3±9.5	+12,1±5.3	+5,28±0.98	+3,0±1.2
SIMMENTALE	72	+0,20±	+13,0±9.5	+ 6,2±5.3	+5,56±0.97	+7,0±1.2
SOUTH DEVON	76	+0,19±	+12,5±9.4	+ 5,0±5.3	+3,47±0.95	+4,0±1.2
HEREFORD (U.K.)	85	+0,07±	+ 4,0±9.3	+ 3,3±5.3	+1,77±0.94	+3,2±1.2
HEREFORD (U.S.)	90	+0,02±	- 0,7±9.4	+ 2,3±5.3	+1,49±0.97	+2,3±1.2
FRIESIAN (N.R.S)		1,25 PT	24,7	1,5	38,91	283,9 D
SEX (♂-♀)	712/690	+0.22±0.05	+15.5±3.6	+5.7±2.0	+2.91±0.49	+1.3±0.4
YEAR (1977-1978)........	732/710	-0.07±0.4	- 4.4±3.2	-2.3±1.8	+0.63±0.32	-1.8±0.4
REGION (Centre-S.W.)....	433/969	-0.24±0.35	-21.0±3.5	-2.8±1.5	-0.85±0.35	+0.8±0.3
Residual standard deviation	1402	0.56 pt	41.6%	23.4%	5.30 kg	4.6 d

-Adusted for dam parity.

FIGURE 2 Sire breed effects on birth weight and frequency of difficult births of crossbreed calves, out of mature Friesian cows (EEC-INRA experiment)

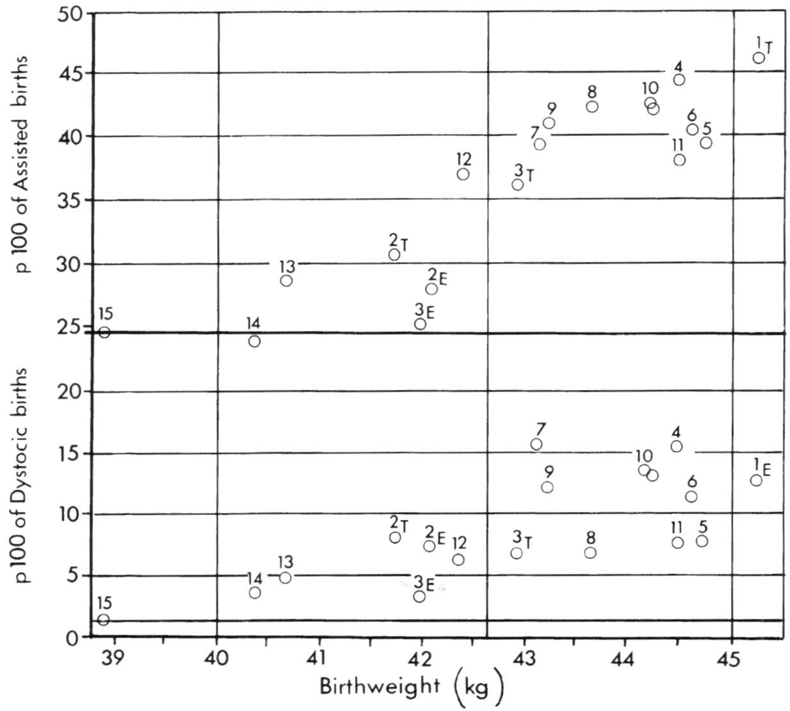

Sire breeds

1 = Charolais
2 = Blond d'Aquitaine
3 = Limousin
4 = Chianina
5 = "Coopelso 93"
6 = "INRA 95"(CH,BA)
7 = INRA 95 (MA,BA,LI)
8 = Blanc Blue Belge
9 = Piemontese
10 = Maine-Anjou
11 = French Simmental
12 = South Devon
13 = Friesian (NRS type)
14 = For crossing type
15 = For purebreeding type

for adult Friesian cows is located around 42 kg; beyond that weight threshold, the frequency of dystocic or assisted births increases very fast with increasing birth weight of the calves, leading to a close correlation between the weight of the calves and their birth difficulties. The average birth weight of the crossbred calves (43.2 kg) tends to be over that threshold and therefore increases the birth difficulties. We generally observe substantial differences between sire breeds or at least between groups of breeds (Figure 2). However there is no significant difference in the magnitude of birth difficulties or their components between all breeds for T.C. and all breeds for P.B. (not including the British beef breeds).

(2) Crossing with British Beef Breeds (HE_{US}, HE_{UK}, SD) noticeably increases the birth weight of the calves (+1.5 to +3.5 kg) although they are small sized breeds. Accordingly, the birth weight of these crossbred calves (41.2 kg) approaches the threshold effect area leading to a slight increment of birth difficulties (+3.5 p.100 of dystocic births). The gestation length generally increases (+3.2 days). In this group, the SD sire breed has produced calves which tend to be heavier (+1.7 kg) than the crossbred HE_{UK} calves and whose birth difficulties exceed those of Friesian calves (+5.0 p.100 of dystocic births). On the other hand, there does not seem to be any marked difference between the two HE types (HE_{US} vs. HE_{UK}) used in this comparison; both are increasing the gestation length (+2.8 days) and birth weight of the calves (+1.6 kg) without substantially enhancing the birth difficulties (+1.7 and +2.8 p.100 assisted and dystocic births, respectively).

(3) Continental beef breeds (CH, BA, LI, CI) definitely increase the gestation length (+6.8 days) and the birth weight (+4.3 kg) of the crossbred calves compared to Friesian calves, producing generally heavier crossbred calves than those of the British beef breeds. For that reason the resulting increase in birth difficulties is

also higher (+8.2 and +11.6 p.100 assisted and dystocic births respectively). However, within this group there are differences between sire breeds. Crossbred CH calves exhibit birth weight similar to that of CI crossbred, and these two sire breeds belong to those inducing most birth difficulties in industrial crossing. Furthermore, BA and LI sire breeds differ little from each other both with respect to birth weight of the calves and the magnitude of birth difficulties; however, the LI sire breed tends to cause less difficulties than the BA sire breed. These two sire breeds (BA and LI) produce much lighter calves (-2.5 kg) than the CH and CI sire breeds and this explains their more restricted incidence of birth difficulties (-7.5 p.100 of dystocic births). As a matter of fact, the birth weight of crossbred LI and BA calves is located almost at the level of the threshold area of birth difficulties (41.4 kg). It should also be noticed that compared with the largest sized British breed (SD), the BA and LI sire breeds have produced crossbred calves of the same mean weight and similar birth difficulties and even less difficult (-6.9 p.100 of assisted births) in particular if considering the LI for P.B. type; on the other hand, their gestation length is significantly longer (+3.5 days). Let us mention that these Continental beef sire breeds especially CI, BA and LI, are those exhibiting the longest gestation length (+7.5 days for BA_T, BA_E, LI_T, LI_E, CI vs FR). Some differences (non significant) appear in the French beef sire breeds between the T.C. type and the P.B. type, and this may reflect differences in the methods used to sample bulls of these two types.

(4) Dual purpose large sized breeds (MA, SI, PI, BB) crossed with Friesian cows also lead to a significant increase in the birth weight of calves (+5.0 kg) and accordingly more birth difficulties (+8.7 p.100 of dystocic births). They may be ranked among continental beef breeds and over the British beef breeds. In this group of breeds, we observe less between breed differences than previously, except with respect to gestation length which in crossbred

PI and SI calves is significantly higher than in crossbred MA and BB calves (4.6 days), the former having a gestation length similar to that of crossbred LI, BA and CH and the latter similar to that of crossbred British beef breeds. The crossbred calves of these four dual purpose sire breeds exhibit almost the same birth difficulties (except perhaps the proportion of assisted births), despite some small differences in the birth weights. These dual purpose sire breeds and especially MA, SI and BB are rather similar to large sized continental beef sire breeds (CH_T, CH_E, CI) than to smaller sized continental beef breeds (LI_T, LI_E, BA_T, BA_E) both in terms of birth weight and difficulties of calving; there is even a trend for crossbred PI calves to be born heavier (+1.3 kg) and with more difficulties than crossbred BA (+4.6 p.100 dystocic births). In those dual purpose sire breeds, there does not seem to be special differences between breeds possessing the double muscle trait (BB, PI) and the others (MA, SI).

(5) Specialised sire lines for crossbreeding (C93, 195_{CB}, 195_{MB}) produce significantly heavier calves (+5.3 kg) and exhibit more difficulties at birth (+10.2 p.100 of dystocic births) than Friesian calves. These results indicate that their birth conditions are located between those of large sized dual purpose sire breeds and those of large sized continental beef sire breeds (CH, CI in particular). Besides, their performance is located between those of the parental breeds of these synthetic lines, but the birth weight of their calves is significantly higher (C93, 195_{CB}). It should be noticed that the double muscled sire lines (195) are rather close to the two dual purpose breeds possessing the double muscle trait (BB, PI). The double muscle line 195_{CB} selected for a longer time, tends to produce a little heavier crossbred calves than those of the PI (+1.4 kg) and BB (+0.9 kg) sire breeds without any increase in the frequency of dystocic births. Likewise there are no differences in the birth conditions of this 195_{CB} line and the C93 line of the same genetic

type.

All these differences between sire breeds and groups of sire breeds have been observed from information supplied by the farmers. Despite the care taken in gathering this data, it is possible that the judgement of the breeder is not fully independent of the aspect of the calf born, especially its coat (colour and marker traits).

Growth of Calves in the Nursery

On the basis of the testings made on male and female calves during their nursery period (2-3 weeks to about 5 months) we compared the sire breeds as they influence the growth and conformation of their calves during the period before weaning from 20 to 120 days. These data were corrected for the same effects as those concerning the birth conditions (Table 2).

(1) Crossbred calves generally exhibited a lower pre-weaning growth (-30g/day) than Friesian calves. The crossbred calves entered the nursery at a higher weight (+4.6 kg) than the Friesian ones, but at 120 days their weights are similar (+1.9 kg). This lower growth rate is even more marked if considering only calves from breeds of the T.C. type (-44g/day); they entered the nursery at an even higher weight (+5.7 kg) but also showed the same weight at 120 days (+1.5 kg). Contrary to methods used in veal production system, the dietary regime of these nurseries were pre-established irrespective of the growth potential and weight of the calves, so as to obtain a restricted growth (708 g/day). At 4 months, the beef conformation and muscling score of crossbred calves give higher scores (+1.2 and +1.0 points, respectively). However, this difference mainly depends on the higher conformation scores recorded in calves from sire breeds of T.C. type (+2.3 and 1.9 points, respectively) whereas calves of the other sire breeds of the P.B. type, be they large sized (-0.8 and +0.6 points, respectively) or British type (-1.1 and -0.6 points, resptectively), gave similar or lower beef conformation and muscling scores than Friesian calves. A certain between

TABLE 2: Sire Breed Effect on Nursery Performance of Calves out Mature Friesian Cows (figures are expressed in deviation to the friesian effect).

	SIRE BREEDS OR MAIN EFFECTS :	NUMBER	INITIAL WEIGHT (kg)	GROWTH RATE (g/day)	120 DAYS WEIGHT: (kg)	BEEF CONFORMATION : (POINTS) TOTAL SCORE :	MUSCLING SCORE :
"TERMINAL SIRE BREEDS" :	CHAROLAISE	80	+6,4±1.4	-19±34	+ 5,0±	+1,5±0.9	+2,5±0.9
	BLOND D'AQUITAINE	71	+3,8±1.4	-55±34	- 1,5±	+1,8±0.9	+0,6±0.9
	LIMOUSINE	71	+5,2±1.4	-78±34	- 3,1±	+2,4±0.9	+1,3±0.9
	"COOPELSO 93"	72	+6,1±1.4	-42±34	+ 2,8±	+2,0±0.9	+1,0±0.9
	"INRA 95" (BA,CH)	68	+6,9±1.4	-27±35	+ 4,9±	+2,9±0.9	+2,8±0.9
	"INRA 95" (MA,BA,LI)....	89	+6,6±1.4	-31±33	+ 2,8±	+2,4±0.9	+2,1±0.8
	BLANC BLEU BELGE	61	+4,8±1.4	-37±35	+ 2,0±	+3,3±0.9	+3,8±0.9
	PIEMONTESE	60	+5,7±1.4	-65±35	- 0,9±	+2,4±0.9	+0,6±0.9
"NON TERMINAL SIRE BREEDS"	CHAROLAISE	59	+7,0±1.5	+32±35	+10,6±	+1,3±0.9	+2,2±0.9
	BLOND D'AQUITAINE	76	+2,8±1.4	-66±34	- 3,7±	+1,4±0.9	+0,2±0.9
	LIMOUSINE	71	+2,9±1.4	-70±34	- 4,5±	+1,8±0.9	+0,4±0.9
	CHIANINA	75	+5,7±1.4	-27±34	+ 3,5±	-1,1±0.9	-2,6±0.9
	MAINE-ANJOU	66	+5,4±1.4	+22±35	+ 8,1±	+0,9±0.9	+2,0±0.9
	SIMMENTALE	70	+5,0±1.4	+15±34	+ 7,0±	+0,4±0.9	+1,2±0.9
	SOUTH DEVON	68	+3,2±1.4	+16±34	+ 4,5±	-1,0±0.9	-0,8±0.9
	HEREFORD (U.K.)	81	+1,7±1.4	-21±34	- 1,2±	-1,2±0.9	-0,3±0.8
	HEREFORD (U.S.)	73	-0,5±1.4	-48±34	- 4,9±	-1,1±0.9	-0,7±0.9
	FRIESIAN (N.R.S.)	67	48,4 KG	743G/D	122,6 KG	22,5 PTS	11,5 PTS
	SEX (O O)	654/625	+3.3±0.6	+46±13	+ 8.1±1.4	-0.9±0.3	+1.1±0.3
	YEAR (1977-1978)	698/580	-0.5±0.5	+34±13	+ 3.5±1.3	+0.0±0.3	+0.8±0.3
	NURSERY (Centre-S.W.) ..	480/798	+0.8±0.6	+13±12	+ 3.4±1.4	-3.0±0.3	-4.3±0.3
	Residual standard deviation	1278	6.8 kg	146 g/d	16.2 kg	4.0 pts	3.7 pts

- adjusted for dam parity.

Sire breed variability within these groups was recorded.

(2) Crossbred calves of British beef breeds (SD, HE_{UK}, HE_{US}) showed almost the same growth performance as Friesian calves in the case of SD and HE_{UK}(+19 g/day), but lower performance in the case of crossbred HE_{US}(-48 g/day). The latter entered the nursery at the same weight as the Friesian calves (-0.6 kg) but reached a lower weight at 120 days (-4.9 kg). Furthermore, the growth of SD crossbred tends to exceed that of HE_{UK} crossbreds (+37 g/day - non significant) and thus the weight of the former calves was higher at 120 days (+5.7 kg). On the other hand, in this group of British beef sire breeds, the beef conformation and muscling score of the calves at 4 months were comparable whatever the sire breed and rather lower than those of Friesian calves (-1.1 and -0.6 points, respectively).

(3) Crossbred calves of continental beef breeds (CH, BA, LI, CI) showed the poorest growth performance in the nursery (-41 g/day) as compared to Friesian calves). Thus, in spite of their higher weight when entering the nursery (+4.8 kg) their weight at 120 days was the same as that of Friesian calves (+0.9 kg) or calves of British sire breeds (+0.3). However characteristic trends can be observed in this group of continental beef sire breeds, especially in the French breeds. Considering crossbred CH calves issued from the 2 types (CH_T,CH_E), these calves exhibited a growth comparable to that of Friesian calves (+6 g/day) and they therefore maintained at 120 days (+7.8 kg) their weight superiority obtained before they entered the nursery (+6.7 kg). This was much more marked for crossbred CH_E reaching the highest weight at 120 days during our trial (133.2 kg). Contrary to that crossbred, BA and LI calves showed the poorest growth performance (-67 g/day as compared to Friesian calves) whatever the type, T.C. or P.B. Thus, although they entered the nursery with a medium weight (+3.7 kg), their weight at 120 days was lower than that of Friesian calves (-3.2 kg)

and comparable to that of HE_{US} (+0.8 kg). At the present state of our analysis it is difficult to impute this astonishing result to a particular pheonomenon (feeding behaviour, feed intake capacity, etc...). During this period, CI crossbred calves tend to be located between the others or to be closer to CH crossbreds. Despite these growth problems limiting the expression of the conformation, the beef conformation (+1.3 points) and muscling score (+0.7 points) of crossbred calves from continental beef breeds were superior to those of the Friesian calves; only CI crossbred calves were highly inferior to Friesian calves, especially with respect to muscling (-1.1 and -2.6 points, respectively). Let us mention that as regards beef conformation, French beef sire breeds can be ranged in the usual order whereas muscling is difficult to estimate in these conditions. In addition although the differences are non significant, the crossbred calves of these sire breeds issued from the T.C. type seem to exhibit a better conformation (+0.4 and 0.5 points, respectively) than those issued from the P.B. type.

(4) Crossbred calves from large sized dual purpose breeds (MA, SI,BB,PI) showed almost the same performances as those of the crossbreds of continental beef sire breeds, but their growth during the nursery period was higher (+24g/day). As compared to Friesian calves, their growth was only a little lower (-16g/day) and they therefore maintained their higher weight reached before entering the nursery (+5.2 kg); their beef conformation (+1.2 points) and especially their muscling score (+1.9 points) were higher. Large differences were observed between these sire breeds and in particular between these sire breeds and in particular between those possessing the double muscle character (BB,PI) and the others (MA, SI). The former(BB, PI) tended to behave as French beef breeds, especially those of the T.C. type (CH_T, BA_T, LI_T) with rather poor growth performance leading to a weight at 120 days similar to that of Friesian

calves, but with a definitely better beef conformation. Conversely the latter (MA, SI), resumbling the CH_E, exhibited higher growth performance than those of Friesian calves. The calves at 120 days were heavy, but the beef conformation tended to be poorer than that of the previous ones, although their better growth performance were favourable to the expression of their conformation.

(5) Crossbred calves from specialised sire lines (C93, 195_{CB}, 195_{MB}) showed almost the same performance in the nursery as those of the crossbred calves of the other sire breeds of the T.C. type, either French beef breeds or dual purpose breeds possessing the double muscle trait. However, at 4 months the weight and beef conformation of these crossbred calves were higher, especially in those belonging to the double muscle sire line 195_{CE}. Let us mention that the crossbred calves of double muscle sire line (195_{CB}) behaved like BB crossbreds but tended to be heavier and to have a better muscling, at 4 months, than PI crossbred calves.

Consequently during this rearing and weaning period of the calves, differences between the sire breeds or groups of sire breeds occur which may affect the subsequent performance of the crossbred calves. A comprehensive study of these differences should be made, in particular in order to find a better adaptation of the dietary regime of crossbred calves before weaning to each different groups.

Fattening Performance of Young Bulls

Our analysis almost exclusively deals with young bulls intensively fattended in performance testing stations and fed ad libitum; some information about the other young bulls has been analysed. For comparing the different sire breeds we analysed the growth performance obtained at the station (at constant age and for the same period) as well as the final weight of the young bulls. Individual testing of the feed intake allowed us to determine their appetite (kg feed intake/100 kg mean live weight) and their feed conversion ratio (kg feed intake/kg weight gain).

The feed contained 88 p.100 dry matter, 0.69 U.F.V. energy and 92g digestible matter protein per kg feed. All these data were corrected for effects of the year, region of origin, fattening station (location x housing conditions) and of the age. Data from the young bulls fattened from 8-14 months and those fattened from 8-17 months were analysed separately. Only the average value of these two estimates is given here (Table 3) in order to reduce the variability due to the sampling between these two groups; these findings therefore correspond to a station fattening period beginning at the age of 243 days (242 and 244 days) and lasting until 481 days (431 and 531 days) for an average fattening length of 238 days (189 and 287 days).

(1) Performance of the crossbred young bulls was better than that of the FR bulls during fattening. At 8 months at the beginning of fattening, the mean weight of the crossbred bulls (+1 kg) was almost the same as that of the FR bulls, but there were differences between the genetic types. During the post-weaning period before fattening from 5 to 8 months, the growth performance (1007g/day) did not much alter the differences observed at 4 months. However, the weight of a large number of crossbred bulls exceeded that of the FR bulls. On the other hand during the intensive fattening, the growth rate of the crossbred young bulls was mostly higher than that of the FR bulls (+70g/day). Accordingly the final weight of the crossbred bulls at 15.1/2 months exceeded that of the FR young bulls; a certain variability between genetic types was observed once again. This higher growth performance in the crossbred young bulls was obtained with a systematically lower feed intake (-0.8kg/day) due to a lower appetite (-0.27kg) especially in the group fattened up to 17 months. The mean effect on their feed conversion ratio was obviously very positive (-1.30). This favourable effect should be related to the better growth rate of crossbred young bulls for the same weight at the beginning of fattening, but should also be associated with the generally later maturity of the sire breeds compared in this trial.

TABLE 3: SIRE BREED EFFECT ON FATTENING PERFORMANCE OF BULLS OUT MATURE FRIESIAN COWS
(figures are expressed in deviation to the Friesian effect).

SIRE BREEDS OR MAIN EFFECTS :	NUMBER	INITIAL WEIGHT (KG)	GROWTH RATE (G DAY)	FINAL WEIGHT : (B) (KG DAY)	FEED INTAKE (KG DAY)	FOOD CONVERSION (FEED GAIN)	APPETITE: (100xFEED LIVEWEIGHT)
CHAROLAISE	16	+19,8±14.5	+114±71	+44,1±21.4	-0,28±0.60	-1,22±0.61	-0,31±0.13
BLOND D'AQUITAINE	15	- 2,4±15.0	+ 33±74	+ 2,3±22.1	-1,29±0.61	-1,34±0.63	-0,35±0.13
LIMOUSINE	16	- 4,8±14.4	+ 8±71	- 2,2±21.4	-1,25±0.59	-1,17±0.61	-0,30±0.13
"COOPELSO 93"	17	- 2,7±14.4	+ 32±71	+ 7,5±21.3	-1,36±0.59	-1,56±0.60	-0,39±0.13
"INRA 95" (BA,CH)	16	+ 3,5±14.6	+101±72	+27,1±21.6	-0,89±0.60	-1,58±0.61	-0,34±0.13
"INRA 95" (MA,BA,LI) ..	17	+10,7±14.4	+ 88±71	+28,4±21.3	-0,72±0.59	-1,25±0.61	-0,34±0.13
BLANC BLEU BELGE	17	+10,4±14.3	+ 71±71	+27,8±21.2	-0,69±0.59	-1,26±0.60	-0,33±0.13
PIEMONTESE	17	-10,8±14.3	+ 64±71	+ 3,2±21.2	-1,26±0.59	-1,67±0.60	-0,31±0.13
CHAROLAISE	17	+17,8±14.5	+113±72	+42,9±21.5	-0,35±0.60	-1,26±0.61	-0,31±0.13
BLONDE D'AQUITAINE	16	- 9,9±14.6	+ 90±72	+10,9±21.7	-0,56±0.60	-1,26±0.62	-0,17±0.13
LIMOUSINE	15	-10,7±14.8	+ 66±73	+ 3,6±21.9	-1,29±0.61	-1,66±0.62	-0,31±0.13
CHIANINA	18	+ 4,5±14.2	+ 92±70	+25,6±21.0	-0,39±0.58	-1,10±0.60	-0,21±0.13
MAINE-ANJOU	18	+13,6±14.1	+ 89±68	+32,9±20.9	-0,03±0.58	-0,79±0.59	-0,17±0.13
SIMMENTALE	17	+ 6,6±14.4	+ 77±71	+23,3±21.4	-0,63±0.60	-1,23±0.61	-0,28±0.13
SOUTH DEVON	16	+ 3,6±14.6	+ 80±72	+21,2±21.8	-0,38±0.60	-1,08±0.62	-0,20±0.13
HEREFORD (U.K.)	17	-12,3±14.4	+ 83±71	+ 7,0±21.4	-0,61±0.59	-1,26±0.61	-0,15±0.13
HEREFORD (U.S.)	17	-20,2±14.4	- 12±71	-23,4±21.4	-1,59±0.59	-1,43±0.61	-0,26±0.13
FRIESIAN (N.P.S.)	15	245,9 KG	1106 D	508,8 KG	11,35 KG D	10,34	3,02
Year (1977/78-1978/79)	149/148	+ 0.8±4.8	+ 60±24	+17.3±7.0	-0.59±0.20	-1.07±0.20	-0.22±0.04
REGION (Centre S.W.)	110/187	- 5.4±4.8	+ 6±24	- 4.4±7.1	+0.09±0.20	+0.05±0.21	+0.06±0.04
Station (Limoges-Soual)	168/60	-12.9±5.8	-117±28	-38.6±8.5	+0.17±0.24	+1.04±0.25	+0.23±0.05
(Tied-box)	60/69	+ 5.9±7.0	+ 99±35	+27.0±10.4	+0.30±0.29	-0.41±0.31	-0.04±0.06
Residual standard deviation	297	25.8 kg	127 g/d	38.2 kg	1.06 kg/d	1.14	0.24

- adjusted for start age; - young bulls fed ad libitum;
(a) at 238 days old; (b) constant final age of 481 days.

(2) Crossbred young bulls of British beef breeds (SD, HE_{UK}, HE HE_{US}) showed higher fattening performance than the FR, except the HE_{UK} crossbred which were the only ones to finish fattening with a lower weight than the FR bulls (-23kg) and whose growth rate (-12g/day) did not compensate for their lower initial weight (-20kg). The young bulls of the two other sire genetic types (SD, HE_{UK}) started fattening with a little lower weight (-8kg) but exhibited a higher growth rate (+82g/day) than the FR bulls and reached a higher weight at the end of fattening (+14kg), especially in the case of crossbreds sired from SD. The appetite of these 3 genetic types was lower than that of the FR bulls (-0.20kg) and their feed conversion ratio was more favourable (-1.26).

(3) Crossbred young bulls from continental beef sire breeds (CH, BA, LI, CI) belong to the groups exhibiting the highest growth performance, but in which also the differences between within-group genetic types are larger. As compared to FR young bulls, their mean weight at beginning of fattening was comparable (+2kg) but their growth rate became thereafter higher (+74g/day), so that their final weight was higher (+18kg). Their appetite during fattening was lower (-0.28kg) and their feed conversion ratio more reduced (-1.27). Although the growth of these young bulls was comparable to that of the crossbreds of the British sire breeds (SD, HE_{UK}), it was obtained with a lower feed intake (-0.28kg). There were marked differences in this group of continental sire breeds, especially between the French beef breeds. As compared to FR, the weight of crossbred CH at the end of fattening was the highest of all the genetic types (+44kg) due to a higher initial weight (+18kg) and growth rate (+113g/day). This higher growth rate did not result from a different appetite in the other crossbred young bulls as compared to the FR bulls (-0.31kg). Let us mention that the results were the same for the 2 types of CH despite the different choice of sires. Conversely, in this group

the crossbred LI and BA bulls exhibited a lower initial weight as compared to the FR (-7kg) and their growth rate (+49g/day) only led to a little higher final weight compared to FR bulls, but a little lower than the crossbreds of British beef sire breeds (SD, HE_{UK}) (-11kg). Compared to the latter, their appetite was lower (-0.11kg) and they had a more favourable feed conversion ratio (-0.19). It is possible that in these continental beef sire breeds (BA, LI) the growth handicap during the nursery period and before fattening has affected the fattening performance. However, crossbred progeny from sires of the P.B. type tended to show better performance especially than BA. The performances of crossbred CI bulls tended to approach those of the crossbred CH before fattening, but they remain lower; thus their weight at the end of fattening is -17kg lower (non significant) than that of crossbred CH_E.

(4) Crossbred young bulls from large sized dual purpose sire breeds (MA, SI, BB, PI) generally showed fattening performance comparable to the mean performance of crossbred bulls from continental beef sire breeds.

As compared to the FR bulls their weight at the beginning of fattening is higher (+5kg) and their subsequent growth was greater (+75g/day), providing them with the largest mean weight at the end of fattening (+22kg). Like the crossbreds of continental beef sire breeds, their appetite was lower (-0.27kg) and their feed conversion ratio more favourable (-1.24). However in comparison with crossbred CH (T.C. and P.B.) the crossbred young bulls of dual purpose sire breeds exhibited lower weights at the beginning of fattening (-14kg) and lower growth rates (-38g/day) leading to a lighter weight at the end of fattening (-22kg), but without any marked difference as regards the appetite (+I.04kg) and feed conversion ratio (-0.C0). In this group of dual purpose sire breeds, PI crossbred bulls showed a lower final fattening weight than the other crossbreds (-25kg); this was more due to a lower weight at the beginning of fattening (-21kg) than to a smaller growth rate (-15g/day). The fattening performance of these crossbred bulls are compared to crossbred LI bulls (P.B. and T.C.). Contrary to the PI crossbreds, the crossbred bulls of the other dual purpose sire breed possessing the double muscle trait (BB) showed analogous performance to those of the crossbreds of the 2 other dual purpose sire breeds (MA, SI) but their appetite was probably smaller and their feed efficiency better. There were only very small differences between crossbred MA bulls and crossbred SI bulls, the former showing a slightly higher growth potential.

(5) Crossbred young bulls from specialised male sire lines (C93, 195_{CB}, 195_{MB}) exhibited fattening performances rather close to the mean performance of crossbred bulls of continental beef sire breeds and large size dual purpose sire breeds, without being significantly superior to the mean performance of the young bulls of their parental breeds. As compared to FH bulls, their weight at the beginning of fattening was slightly higher (+4kg) and their growth rate better (+74g/day) so that they were the heaviest young bulls at the end of fattening (+21kg). These performances were obtained with the lowest feed intake (-1.0kg/day) due to their lower appetite (-0.36kg) and one of the best feed conversion ratios (-1.46). In this group, the young bulls from the two 195 double muscle sire lines (195_{CP}, 195_{MB}) seemed to have higher growth performance (+7kg weight at the beginning of the fattening; +95g/day weight gain per day and +28kg weight at the end of the fattening) and their feed conversion ratio was efficient (-1.42). This performance was comparable to those of BB crossbred bulls but lower than of PI crossbred bulls.

(6) The role of these sire breeds or genetic types in industrial crossing in improving the growth potential of young bulls, partly depends on their fattening system. Indeed, it has been shown that the better efficiency of

crossbred young bulls during intensive fattening period primarily depends on an improvement of the growth rate obtained reducing the appetite comparatively to FR. When reducing the feed intake in the same way in all the group bulls studied (according to the metabolic weight) we observed a marked increase in their feed efficiency. Being analysed according to a model with fixed effects and without sire effects, the differences between genetic types are difficult to interpret because of the small number of bulls and the interferences with performance before the fattening period. Generally (Table 4) we reduced the feed intake of the bulls by 14-15 p.100, i.e., much more than initially planned (21 instead of 10-15 p.100 for FR up to 14 months; 15 instead of 7-8 p.100 for crossbred CH up to 17 months). The appetite of young bulls belonging to later maturing genetic types from 8-17 months was thus proportionately less reduced (-13.8 p.100) than that of those sired from earlier maturing genetic types, between 8-14 months (-15.6 p.100). These differences were even larger if considering the extremes of these two cases, i.e. crossbreds of specialised sire lines (-12.2 p.100) and FR (-17.3 p.100). To compensate for the growth reduction due to feed restriction, the young bulls finished their fattening period 14 days later than the others not restricted. The feed restriction led to a proportionally larger growth reduction in later maturing young bulls at 17 months (-47g/day or -4.1 p.100) than in those of the earlier maturing type at 14 months (-12g/day or -1.0 p.100). Accordingly, the feed conversion ratio of early maturing bulls was much more improved (-14.6 p.100) than that of later maturing bulls (-10.1 p.100) and especially crossbred young bulls of continental beef sire breeds (-9 p.100) or of specialised sire lines (-9.5 p.100). This seems to indicate that the better efficiency of restricted maturing bulls results above all from a change in the composition of their weight gain (leaner). These results which are not called in question by the fattening performance of young bulls fed with

Table 4: Effect of Feed Intake Restriction (A) on Fattening Performance of Friesian Crossbred Bulls Sired by Different Types of Cattle Breeds

SIRE BREEDS :	Number (ad lib. /restr.)	FINAL AGE:	INITIAL WEIGHT:	GROWTH RATE:	FINAL WEIGHT :	FEED INTAKE :	FEED CONVERTION (Feed/ growth) :	APPETITE (Feed/live weight):
*SPECIALISED MALE LINES (C93, 195 CR, 195 MB)	60/52	17 months	(b) 251.1kg (c) +0,4(0,2)	1161 g/day +38(3,3)	583.6 kg -0,2(-0,0)	10.84kg/day +1,24(+11,5)	9.44 +0,90(+9,5)	2.60p.100 +0,32(+12,2)
*CONTINENTAL BEEF BREEDS (CH, BA, LI, CI)	26/22	17 months	251.1kg -1,1(+0,4)	1142 g/day +54(+4,8)	578.0 kg +3,2(+0,6)	10.96kg/day +1,46(+13,3)	9.63 +0,87(+9,0)	2.65p.100 +0,35(+13,3)
*LARGE SIZED DUAL PURPOSE BREEDS (MA, SI, BR, PI)	38/27	17 months	258.8kg +1,1(+0,4)	1152 g/day +41(+3,6)	588.2 kg +1,3(+0,2)	11.31 kg/day +1,76(+15,6)	9.90 +1,22(+12,3)	2.71p.100 +0,43(+15,7)
*LATE MATURING BREEDS	124/101	17 months	253,3 KG -0.2(-0.1)	1149 G/DAY 47(+4,1)	5[illegible]2,1 KG 1,9(+0,3)	11,05KG/DAY 1,50(+13,6)	9,67 0,97(+10,1)	[illegible] 0,37(+13,8)
*BRITISH BEEF BREEDS (SD, HE UK, HE US)	28/22	14 mo.	235.8 kg -- 3.0(-1,3)	1190 g/day +12(+1,0)	459.7 kg -15,9(-3,5)	10.16 kg/day +1,27(+12,5)	8.66 +1,04(+12,0)	2.95p.100 +0,44(+15,1)
*FRIESIAN (NRS)	8/8	14 mo.	243.2 kg +21,9(+9,0)	1127 g/day +13(+1,2)	455.6 kg +9,3(+2,9)	10.86 kg/day +2,32(+21,4)	9.86 +2,11(+21,4)	3.12p.100 +0,54(+17,3)
*EARLY MATURING BREEDS	36/30	14 mo.	237,6KG +3,2(1,4)	1174 G/DAY +12(1,0)	458,7 KG -9,2(-2,1)	10,34 KG/DAY +1,54(14,8)	8,90 +1,31(14,0)	2,99P.100 +0,47(15,6)

(a) : Restriction proportional to metabolic weight

(b) : Average performances of bulls fed ad libitum.

(c) : Difference between ad libitum and resticted (relative value in p.100).

silage, in excess; enhance the interest of adapting the mode and level of feeding to the growth potential and to the ingestion capacity of the different genetic types of young bulls.

Weight and Morphology of Young Bull Carcasses

Like the results obtained in the fattening trial, the slaughter performance of the young bulls was analysed and interpreted with the same model. The mean slaughter performances at constant age is given in Table 5. On average, these *ad libitum* fed young bulls were slaughtered at the age of 16 months and at a similar weight of 508.3kg when arriving at the slaughterhouse (460.4kg and 566.1kg, respectively at 14 and 18 months). The dressing percentage corresponding to the ratio between the cold carcass weight and the live weight at the arrival at slaughterhouse after transport (about 10 or 300 km according to the station). The morphology of the carcasses was estimated either by measurement or weight ratios according for their compactness at various levels or by a fleshiness score according to the E.A.A.P. notation (1 to 5 classes and 3 sub-classes - DE BOER et al., 1974).

(1) All crossbred young bulls showed a much heavier carcass weight than FR bulls (+21.5kg) resulting both from a higher live weight at slaughtering (+14.5kg) and from a higher dressing percentage (+2.3 p.100). In addition their carcasses were more compact whatever the region considered and showed a greater fleshiness (+1.6 points i.e. about 1/2 classes). Beside this superiority of crossbred young bulls, there were also a large variability between sire breeds.

(2) The crossbred young bulls of British beef sire breeds (SD, HE_{UK}, HE_{US}) also exhibiting the same slaughter weight as FR bulls (-1.3kg), give carcasses with a substantially higher weight (+4.7 kg) because of their higher dressing percentages (+1.2 p.100). However, there were marked differences between these 3 types of crossbred young bulls as regards the carcass weight especially because of the

TABLE 5: Sire Breed Effect on Slaughtering Performance and Carcass Morphology of Freisian Crossbred Bulls at contant age: (16 months old)

		Number	SLAUGHTER WEIGHT (KG)	DRESSING CUT (P.100)	COLD CARCASS WEIGHT (KG)	CARCASS MORPHOLOGY : LEG THICKNESS / LEG LENGTH (10^2)	LOIN THICKNESS / LOIN LENGTH (10^2)	CARCASS WEIGHT (KG) / CARCASS LENGTH (CM)	FLESHINESS (E.E.A.P. Conformation score)
"TERMINAL CROSSING" :	CHAROLAISE	16	+47,5± 22.5	+3,4±0.9	+40,6±14.0	+2,6±0.8	+2,8±0.7	+3,0±0.9	+2,8±0.8
	BLONDE D'AQUITAINE	15	+ 2,4± 22.5	+3,9±0.9	+20,0±14.4	+2,1±0.8	+2,1±0.8	+1,8±0.9	+2,1±0.8
	LIMOUSINE	16	- 5,2± 21.8	+2,8±0.9	+10,0±14.0	+1,4±0.8	+1,9±0.7	+1,3±0.9	+1,7±0.8
	"COOPELSO 93"	17	+ 4,7± 21.9	+4,5±0.9	+23,3±13.9	+1,8±0.8	+1,9±0.7	+1,8±0.9	+2,2±0.8
	"INRA 95" (BA,CH)	16	+23,3± 22.0	+3,3±0.9	+29,4±14.1	+2,0±0.8	+2,2±0.7	+2,3±0.9	+2,0±0.8
	"INRA 95" (MA,BA,LI) ..	17	+24,8± 21.7	+2,7±0.9	+27,1±13.9	+1,9±0.8	+1,7±0.7	+1,9±0.9	+2,2±0.8
	BLANC BLEU BELGE	17	+21,2± 21.6	+4,7±0.9	+35,3±13.8	+2,2±0.8	+1,7±0.7	+2,5±0.9	+2,5±0.8
	PIEMONTESE	17	+ 6,3± 22.0	+4,8±0.9	+25,0±13.8	+1,2±0.8	+2,2±0.7	+1,9±0.8	+1,6±0.8
"NON TERMINAL CROSSING" :	CHAROLAISE	17	+42,7± 22.2	+2,4±0.9	+30,0±14.0	+2,0±0.8	+1,9±0.7	+2,4±0.9	+2,1±0.8
	BLONDE D'AQUITAINE	16	+ 6,0± 22.1	+3,8±0.9	+21,3±14.2	+1,4±0.8	+1,8±0.7	+1,6±0.9	+1,6±0.8
	LIMOUSINE	15	- 3,5± 23.1	+2,8±0.9	+13,0±14.2	+1,3±0.8	+1,7±0.7	+1,5±0.9	+2,0±0.8
	CHIANINA	18	+20,2± 22.3	+3,2±0.9	+27,6±13.7	+0,8±0.8	+0,5±0.7	+2,1±0.9	+0,7±0.7
	MAINE-ANJOU	18	+31,3± 21.3	+2,2±0.9	+28,2±13.6	+0,7±0.8	+1,4±0.7	+1,8±0.9	+1,0±0.8
	SIMMENTAL	16	+28,4± 22.5	+1,3±0.9	+20,7±14.2	+0,3±0.8	+1,5±0.7	+1,5±0.9	+1,1±0.8
	SOUTH DEVON	16	+ 23,5± 22.7	+1,5±0.9	+18,5±14.2	+0,5±0.8	+0,7±0.7	+1,4±0.9	+0,1±0.8
	HEREFORD (U.K.)	17	+ 0,8± 21.8	+1,1±0.9	+ 5,6±14.0	+2,5±0.9	+1,1±0.7	+0,9±0.9	+0,9±0.8
	HEREFORD (U.S.)	17	-28,1± 22.1	+0,9±0.9	-10,0±13.9	+1,5±0.8	+1,2±0.7	-0,9±0.9	+0,3±0.8
	FRIESIAN (N.R.S)	14	492,2KG	53,9%	266,0KG	35,8P.100	13,7P.100	20,8 KG/CM	3,8 PTS (A)
	Year (1977/78-1978/79)	147/148	+25.8±7.3	+1.3±0.3	+19.8±4.5	+0.9±0.3	+0.8±0.2	+1.5±0.3	+0.8±0.2
	Region (Centre S.W.)	108/186	- 7.2±7.3	+0.4±0.3	- 1.8±4.6	+0.1±0.3	+0.1±0.2	+0.2±0.3	+0.3±0.3
	Station (Limoges/Soual)	166/60	-72.6±8.9	+1.0±0.4	-34.1±5.5	-0.7±0.3	-0.5±0.3	-1.7±0.4	-0.2±0.3
	(Tied-box)	60/69	+30.3±10.9	+0.4±0.4	+18.8±6.7	+1.3±0.4	+0.8±0.4	+0.8±0.4	+0.6±0.4
	Residual standard deviation	295	38.4kg	1.5%	24.5 kg	1.5p.100	1.3p.100	1.6 kg/cm	1.3 pts

- adjusted for constant age. - young bulls fed ad libitum. (a) : equivalent to "3" class.

differences in weight before slaughtering; the final weight of crossbred SD bulls exceeded by +12.9 kg that of HE_{UK} crossbreds whose carcasses weighed +15.6kg more than those of the other crossbred HE_{US}. The fleshiness of the carcasses from crossbred bulls of British beef sire breeds scored only a little higher than that of the FR bulls (+0.4 subclass) whereas their carcasses seemed to be more compact in particular at the level of the back-quarter in HE crossbred.

(3) Crossbred young bulls of continental beef sire breeds (CH, BA, LI, CI) belong to those producing the significantly heaviest carcasses compared to those of the previous types (+23.2kg/FR and +18.5kg/British beef breed crossbreds). This superiority compared to FR bulls, resulted both from a higher slaughter weight (+15.7kg) and especially from a definitely higher dressing percentage (+3.2 p.100). Their carcasses were also more compact and showed a highly improved fleshiness (+1.9 subclass). In this group of crossbreds, CH crossbreds and especially those of the T.C. type had the heaviest carcasses compared to FR bulls (+35.3kg) depending more on their slaughter weight (+45.1kg) than on their dressing percentage (+2.9 p.100). Their carcasses were also more compact and showed a higher fleshiness, about one class higher (+2.5 subclass). CI crossbreds also showing the same dressing percentage as CH crossbreds (+0.3 p.100) had a lower carcass weight (-7.7kg) because of their lower slaughter weight (-24.9kg) and their carcasses showed less compactness and fleshiness (-1.8 subclass). The carcass weight of crossbred BA and LI was lower than that of CH crossbreds, but was still higher than that of the FR bulls (+16.0kg) although they had the same weight before slaughter (-0.1kg). Their large superiority as regards the dressing percentage (+3.3 p.100) explains much of this difference. The carcasses of BA crossbreds showed the same fleshiness but were definitely heavier (+4.6kg) than those of LI crossbreds as their weight (+4.3kg) and dressing percentage

(+0.5 p.100) tended to be higher. No large differences appeared between P.B. and T.C. types in these 2 sire breeds.

(4) Crossbred young bulls of large sized dual purpose sire breeds (MA, SI, BB, PI) showed a significantly higher weight (+21.8kg) and dressing percentage (+3.2 p.100) and produced significantly heavier carcasses than those of FR bulls (+27.3kg). Their carcass weight was comparable (even superior : +4.1kg, because of a heavier slaughter weight : +5.9kg) to that of crossbreds from continental beef sire breeds and showed the same compactness and fleshiness. In this group of dual purpose sire breeds the crossbreds MA and SI gave heavier carcasses (+24.4kg) than because of their dressing percentage (+1.8 p.100). The compactness and fleshiness of their carcasses were lower than those of the crossbreds of continental beef sire breeds but a little higher than those of the cross-breds of the crossbreds of British beef sire breeds. Carcasses of MA crossbreds tended to be heavier than those of SI crossbreds, but their morphology was identical. Conversely, crossbreds of dual purpose sire breeds possessing the double muscle character (BB, PI) exhibited the heaviest carcasses (+30.2kg) more because of their superiority as regards the dressing percentage (+4.7 p.100) than because of their live weight (+13.7kg). Besides their dressing percentage was higher than that of BA and LI crossbreds (+1.4 p.100). The fleshiness and compactness of these crossbreds from double muscle dual purpose sire breeds, were comparable to those of the crossbreds of continental beef sire breeds. The carcasses of BB crossbreds, of identical fleshiness but of higher weight (+10.0kg) than those of PI crossbreds, showed a lower weight than those of CH_T crossbreds of the T.C. type (-5.3 kg) but with the same fleshiness and the same compactness.

(5) Crossbred young bulls of specialised sire lines (C93, 195_{CB}, 195_{MB}) almost showed similar performances to those of crossbred continental beef sire breeds or double muscle

dual purpose sire breeds. Compared to FR bulls, their young bulls were heavier at slaughter (+17.6kg) and showed a better dressing percentage (+3.4 p.100) leading to a markedly higher carcass weight (+26.6kg). The fleshiness and compactness of their carcasses were also among the highest values. The carcass weight of the crossbred double muscle sire lines 195 tended to be higher than those of the C93 crossbreds (+5.5kg). The differences between these two double muscle sire lines (195) were not large. Their performance were a little lower than those of CH_T crossbreds and did not significantly differ from those of double muscle dual purpose sire breeds (except perhaps as regards dressing percentage) in spite of the different methods used for sire sampling.

(6) In young bulls subjected to feed restriction and slaughtered at live weight close to that of the same genetic type fed ad libitum, no significant effect of this feed restriction was observed on carcass weight and dressing percentage (Table 6). A slight trend towards a better dressing percentage could however be noticed. The fleshiness and compactness of their carcasses was not much changed; only the carcasses of early maturing crossbreds seemed to show a little less fleshiness because of the feed restriction (-5 p.100). The effect of limiting the feed consumption in the crossbreds of British beef sire breds was characterised by reduction of live weight at slaughter in the bulls subjected to this feed restriction.

Carcass Composition and Meat Qualities of Young Bulls

We analysed the estimated composition of the carcasses according to the same model as that applied to slaughter performance (Table 7). The separation between fat, bone and muscle allowed us to appreciate on the one hand the adiposity or fatness of the carcasses at constant age (p.100 of fat) and on the other hand the composition of the lean meass of the carcasses (muscle to bone ratio). Let us mention that the estimation of magnitude of the superficial fat layer on the carcasses

Table 6: Effect of Feed Intake Restriction (A) on Slaughtering Performance of Friesian Crossbred Bulls sired by different types of Cattle Breeds

	Number (ad lib /restr.)	Slaughter Age	Slaughter Weight:	Dressing Out	Cold Carcass Weight	Fleshiness : (EAAP conformation score (e)
*SPECIALISED MALE LINES (C93, 195_{CB}, 195_{MB})	26/22	17 MTHS	(b) 563.1 kg (c) -0,0(-0,0)	57.8 p.100 -0,3(-0,6)	325.1 kg -3,4(-1,0)	5.7 pts -0,1(-1,4)
*CONTINENTAL BEEF BREEDS (CH,BA,LI,CI)	60/52		557.3 kg +1,8(+0,3)	57.7 p.100 -0,3(-0,5)	320.2 kg -2,3(-0,7)	5.7 pts -0,0(-0,6)
*LARGE SIZED DUAL PURPOSE BREEDS (MA,SI,BB,PI)	38/27		570.0 kg +4,3(±0,8)	57.9 p.100 -0,1(-0,2)	327.8 kg -0,8(-0,2)	5.6 pts +0,2(+3,3)
*LATE MATURING BREEDS..............	124/22		562,1 KG +2,1(+0,0)	57,8 P,100 -0,3(-0,4)	323,4 KG -2,1(-0,7)	5,7 PTS +0,0(+0,8)
*BRITISH BEEF BREEDS (SD, HE_{UK}, HE_{US})	28/22	14 MTHS	442.5 kg -11,5(-2,6)	54.0 p.100 -0,4(-0,7)	238.9 kg -8,3(-3,5)	4.1 pts -0,2(-4,8)
*FRIESIAN (N.R.S.)................	8/8		441.5 kg +1,2(±0,3)	52.9 p.100 -0,3(-0,6)	233.9 kg -1,2(-0,5)	3.5 pts -0,2(-5,7)
EARLY MATURING BREEDS	36/30		422,2 KG -8,3(-1,9)	53,7 P,100 -0,4(-0,7)	237,7 KG -0,5(-2,7)	3,9 PTS -0,2(-4,9)

(a) : Restriction proportional to metabolic weight.
(b) : Average performance of bulls fed ad libitum.
(c) : Difference between ad libitum and restricted (relative value in p.100).
(d) : 14 days more for restricted bulls.
(e) : 4.0 pts = "3⁻" class.

TABLE 7: Sire Breed Effect on Carcass Composition of Friesian Crossbred Bulls at Constant Age (16 months old)

	SIRE BREEDS OR MAIN EFFECTS :	Number	ESTIMATED CARCASS COMPOSITION FAT (P,100)	LEAN (P,100)	BONE (P,100)	LEAN/BONE	FAT COVERING (E.E.A.P. SCORE)
"TERMINAL CROSSING"	CHAROLAISE	16	-2,1±0.9	+3,0±0.9	-0,8±0.3	+0,5± 0.13	-1,3±0.7
	BLONDE D'AQUITAINE	15	-3,0±0.9	+3,7±1.0	-0,3±0.3	+0,37±0.13	-2,6±0.8
	LIMOUSINE	16	-1,0±0.9	+1,2±0.9	-0,1±0.3	+0,14±0.12	-1,3±0.7
	"COOPELSO 93"	17	-2,8±0.9	+3,5±0.9	-0,2±0.3	+0,28±0.12	-2,4±0.7
	"INRA 95" (BA,CH)......	16	-2,3±0.9	+3,0±0.9	-0,5±0.3	+0,41±0.16	-1,6±0.8
	"INRA 95" (MA,BA,LI) ..	17	-2,1±0.9	+2,8±0.9	+0,0±0.3	+0,20±0.12	-1,5±0.7
	BLANC BLEU BELGE	17	-2,5±0.9	+3,3±0.9	-0,4±0.3	+0,37±0.12	-2,1±0.7
	PIEMONTESE	17	-3,5±0.9	+4,2±0.9	-0,7±0.3	+0,56±0.12	-2,2±0.7
"NON TERMINAL CROSSING"	CHAROLAISE	17	-0,9±0.9	+1,5±0.9	-0,1±0.4	+0,17±0.12	-0,8±0.8
	BLONDE D'AQUITAINE	16	-2,4±0.9	+3,0±0.9	-0,1±0.3	+0,24±0.12	-1,5±0.8
	LIMOUSINE	15	-2,0±0.9	+2,4±1.0	-0,1±0.3	+0,24±0.13	-1,3±0.7
	CHIANINA	18	-1,6±0.9	+2,2±0.9	+0,4±0.3	+0,00±0.12	-1,1±0.7
	MAINE-ANJOU	18	-1,3±0.8	+1,9±0.9	+0,5±0.3	-0,01±0.12	-1,3±0.7
	SIMMENTAL	16	-1,4±0.9	+1,9±1.0	-0,1±0.3	+0,16±0.13	-1,5±0.7
	SOUTH DEVON	16	-0,6±0.9	+1,0±0.9	-0,2±0.3	+0,14±0.12	-0,4±0.8
	HEREFORD (U.K.)........	17	+0,8±0.9	-0,8±0.9	-0,5±0.3	+0,13±0.12	+1,1±0.7
	HEREFORD (U.S.)	17	-0,1±0.9	-0,1±0.9	-0,3±0.3	+0,10±0.12	+0,5±0.7
	FRIESIAN (N.R.S.)	14	16,3P,100	68,3P,100	14,1P,100	4,84 PTS	8,6 PTS (A)
	Year (1978-1979)	147/148	+0.7±0.3	-0.5±0.3	-0.3±0.1	+0.06±0.04	+ 0.3±0.2
	Region (Centre S.W.)...	108/186	+0.1±0.8	-0.1±0.3	+0.0±0.1	-0.02±0.04	+ 0.1±0.2
	Station (Limoges Soual)	166/60	-1.7±0.4	+1.2±0.4	+1.0±0.1	-0.26±0.05	- 1.3±0.3
	(Tied-box)	60/69	+2.7±0.4	-2.5±0.5	-1.0±0.2	+ 0.19±0.06	+ 1.5±0.4
	Residual standard deviation	295	1.6%	1.7%	0.6%	0.22 pts	1.3 pts

according to the E.A.A.P. scoring (De Boer et al., 1974), lead to the same grading of the various genetic types as that based on the fat percentage, likewise the results of the commercial carcass cutting (marketable meat, bone, fat plus offal) are in good relation with the carcass composition that we estimated (r=+0.6 to +0.7, Bonaiti, 1980).

(1) The carcass composition of crossbred young bulls were more favourable than that of pure bred FR bulls. Their carcasses were leaner (-1.7 p.100) at the same age thus confirming their later maturity and explaining part of their better feed efficiency. The lean mass of their carcasses also contained less bone (-0.2 p.100) but in particular more muscle (+2.8 p.100) and accordingly showed a much better muscle/bone ratio than FR bulls (+0.24).

(2) Carcasses of young bulls sired from British beef breeds (SD, HE_{UK}, HE_{US}) were generally as fat as those of FR carcasses (+0.04 p.100). However, SD crossbreds tended to be later maturing than HE corssbreds (-1.4 p.100 of fat/HE_{UK}) whose carcasses were fatter than those of FR (+0.8 p.100). On the other hand the carcasses of the crossbreds of British beef sire breeds showed an improved lean mass compared to the FR (+0.12 muscle/bone) due in particular to a lower proportion of bone (-0.5 p.100).

(3) The carcass composition of crossbred young bulls of continental beef sire breeds (CH, BA, LI, CI) was significantly different from that of FR and of crossbreds from British beef sire breeds. Compared to the FR, their carcasses were leaner (-1.9 p.100) and their lean mass contained much more muscle (+2.4 p.100) and less bone (-0.2 p.100) (+0.25 muscle/bone). In this group of continental beef sire breeds the carcass adiposity is less reduced in the crossbred CH and more particularly in those of the T.C. type (-1.5 p.100); on the other hand their carcasses contained proportionally less bone (-0.5 p.100) and a lean mass with more favourable muscle/bone ratio (+0.33). Crossbred CI whose carcasses tended to be leaner

than those of crossbred CH (-0.2 p.100) contained more bone than FR (+0.4 p.100) and accordingly their muscle/bone ratio was therefore comparable to that of the latter (+0.06). The BA and LI crossbreds exhibited a later maturity than CH crossbreds (-0.6 p.100) and more particularly than those of the P.B. type. Contrary to observations generally made, their lean mass contained as much (P.B. type) or less muscle (T.C. type) than that of crossbred CH. However their muscle/bone ratio remained higher than that of crossbreds from British beef sire breeds.

(4) Carcasses of crossbreds from large sized dual purpose sire breeds (MA, SI, BB, PI) showed generally the same composition as those of crossbreds from continental beef sire breeds. In comparison with the FR, they were leaner (-2.2 p.100) contained less bone (-0.2 p.100) and more muscle (+2.8 p.100) (+0.27 muscle/bone). However there was a marked difference between dual purpose sire breeds possessing the double muscle traits (BB and PI) and the others (MA and SI). For the former, the carcass of the young bulls showed much later maturity (-3.0 p.100/FR) and lean mass with a very much higher proportion of muscle (+3.8 p.100) as well as a lower proportion of bone (-0.5 p.100) (+0.70 muscle/bone); consequently their composition tended to be more favourable at constant age than that of the crossbreds among continental beef sire breds especially PI crossbreds. Conversely, the carcasses of the crossbreds of the other dual purpose sire breeds (MA and SI) were fatter (+1.7 p.100/BB+PI) with a larger bone percentage (+0.7 p.100) and a smaller proportion of muscle (-2.4 p.100) (-0.39 muscle/bone). This seemed to be more marked for MA and SI crossbreds.

(5) Crossbred young bulls from specialised sire lines (C93, 195_{CB}, 195_{MB}) showed on average the most favourable carcass composition : -2.4 p.100 of fat and -0.2 p.100 of bone relative to FR, with +3.1 p.100 of muscle (+0.30 of muscle/bone). The performance of C93 crossbreds were close to those of BA crossbreds. There were little

differences between the two double muscle sire lines but where it tended toward a higher proportion of bone (+0.5 p.100) in those containing MA (195_{MB}). The crossbred young bulls of the other double muscle sire line (195_{CB}) were comparable to those progeny from double muscle dual purpose sire breeds and more particularly from BB breeds.

(6) The physico-chemical qualities of the meat measured on a sample of longissimus dorsi muscle showed little differences between genetic types. We noticed the following trends among the main groups of sire breeds (Table 8): No pH differences between these groups; on the other hand the meat of crossbred young bulls tended to be paler and more tender and to contain less collagen and lipids, and to lose a little more water; the meat qualities of crossbred bulls from specialised sire lines and double muscle dual purpose sire breeds were those differing most from those of FR bulls especially as regards the colour, tenderness and lipid content; the meat qualities of crossbred bulls from British beef sires tended to be similar to those of the FR breed in particular with respect to tenderness and lipid content contrary to continental beef sire breeds.

(7) Feed restriction of young bulls has a significant influence on their carcass composition (Table 9). In restricted bulls, the carcasses were significantly leaner whatever the genetic types. This fat reduction was especially marked in FR young bulls (-1.1 p.100) and in crossbreds of dual purpose sire breeds (-0.8 p.100). The feed restriction also tended to reduce the muscle/bone ratio by decreasing the muscle proportion. This change seemed to be rather equal for all genetic groups. There was no important effect on meat qualities except a smaller loss of water in the crossbreds of specialised sire lines and double muscle dual purpose sire breeds after feed restriction.

Conclusions

All the results of this comparison of sire breeds for industrial

TABLE 8: Effect of Type of Sire Breed on Meat Traits of Friesian Crossbred Bulls at Constant Age (16 Months Old)

Sire Breed Type :	Number	pH	COLOUR: Reflectance (p.100)	COLOUR: Hoeminic Iron (µg/g)	HARDNESS: Collagen Content (p.100)	HARDNESS: Shear Force Value : (kg/cm²)	Water Losses: (p.100)	Intra Muscular Lipid Content: (p.100 of weight)
		(a)	(a)	(b)	(b)	(a)	(a)	(b)
*SPECIALISED MALE LINES ($C93, 195_{CB}, 195_{MB}$).........	50	+0.01	+1.3	-2.29	-0.35	-0.29	+0.9	-0.5
*CONTINENTAL BEEF BREEDS (CH,BA,LI,CI)............	113	+0.10	+1.0	-2.07	-0.24	-0.13	+0.2	-0.4
*LARGE SIZED DUAL PURPOSE BREEDS (MA,SI,BB,PI)......	68	+0.08	-1.0	-1.52	-0.21	-0.08	+0.8	-0.5
(BB+PI)-(MA+SI)		(-0.00)	(+1.0)	(+0.04)	(-0.21)	(-0.42)	(+0.3)	(-0.4)
*BRITISH BEEF BREEDS (SD, HE_{UK}, HE_{US})...........	50	+0.05	+1.5	-0.68	-0.07	-0.04	+0.9	-0.1
*FRIESIANS (NRS)..........	14	5.70	55.8P.100	15.00µG/J	2.52P.100	3.73KG/CM²	16.9P.100	2.6 P.100
Residual standard deviation.................	295	0.24	7.7	1.69	0.30	1.1	2.6	0.57

(a) After refrigeration during 6 days.
(b) After congelation, only on 70 carcasses.

\- young bulls fed ad libitum.

TABLE 9: Effect of Feed Intake Restriction (A) on Carcass Composition of Friesian Crossbred Bulls sired by different types of Cattle Breeds

Sire Breeds	Number (ad lib/ restr).	Slaughter Age (d)	Fat p.100 :	Lean p.100:	Bone p.100 :	Lean/Bone :	Fat Covering (E.E.A.P score)
*SPECIALISED MALE LINES (C93, 195_{CB}, 195_{MB})	26/22	17 MTHS	(b) 14.6p.100 (c) +0.7(+4.6)	71.1 p.100 -0.8(-1.1)	13.5 p.100 +0.1(+1.0)	5.26 -0.11(-2.1)	7.7 pts +0.4(+5.3)
*CONTINENTAL BEEF BREEDS (CH,BA,LI,CI)	60/52	17 MTHS	15.0p.100 +0.7(+4.5)	70.7 p.100 -0.8(-1.1)	13.4 p.100 +0.1(+0.4)	5.27 -0.08(-1.5)	7.8 pts +0.3(+3.2)
*LARGE SIZED DUAL PURPOSE BREEDS (MA,SI,BB,PI)	38/27	17 MTHS	15.0p.100 + 0.8(+5.3)	70.8 p.100 -0.9(-1.2)	13.3 p.100 -0.1(-0.4)	5.31 -0.05(-0.9)	8.1 pts +0.8(+9.5)
LATE MATURING BREEDS	124/101	17 MTHS	14.9p.100 +0.7(+4.8)	70.8 p.100 -0.8(-1.1)	13.4 p.100 +0.0(+0.0)	5.28 -0.08(-1.5)	7.8 PTS +0.4(+5.5)
*BRITISH BEEF BREEDS (SD, HE_{UK}, HE_{US})	28/22	14 MTHS	15.7p.100 +0.4(+2.4)	68.3 p.100 -0.6(-0.9)	14.3 p.100 +0.2(+1.3)	4.78 -0.10(-2.1)	8.4 pts +0.7(+8.9)
FRIESIAN (NRS)	8/8	14 MTHS	16.1p.100 ±1.1(+6.7)	67.9 p.100 -1.2(-1.7)	14.5 p.100 -0.1(-0.4)	4.69 -0.06(-1.3)	8.3 pts ±1.1(+14.8)
EARLY MATURING BREEDS ...		14 MTHS	15.8p.100 +0.6(3.5)	68.2 p.100 -0.8(-1.1)	14.4 p.100 +0.1(0.9)	4.75 -0.09(-1.9)	8.4 PTS +0.8(+9.6)

(a) Restriction proportional to metabolic weight.
(b) Average performance of bulls fed ad libitum.
(c) Difference between ad libitum and restricted (relative value in p.100).
(d) 14 days more for restricted bulls.
(e) 7.0 pts = "3^-" class.

crossing should be synthesised in order to determine their respective overall efficiency. However, the following conclusions may be drawn from this preliminary analysis.

For intensive fattening of young bulls, all sire breeds used in the crossing improve the slaughter value of crossbred bulls as they increase their weight and carcass qualities by reducing their fattening precocity. However, this improvement was obtained after increasing the birth weight of the calves and this leads to a significantly higher percentage of difficult births.

Also no sire breed seems to be systematically superior to the others. There is large variability between sire breeds as regards the slaughter value mainly depending on the differences in their size, muscling and maturity. If industrial crossing with British beef sire breeds (small size, good muscling, higher adiposity) permits one to limit the increase in birth weight of the calves and consequently the frequency of dystocic births, it improves the fattening efficiency of the young bulls without supplying a higher slaughter value. Contrary to that, continental beef sire breeds and their specialised sire lines (large size, high muscling and late maturity) improve much more the fattening efficiency and slaughter value (carcass weight and composition) of the crossbred young bulls but they lead to significantly more difficult births because of their high growth potential. However in this group, it is possible to find sire breeds with a lower growth potential resembling the British beef sire breeds but with higher muscling and later maturity. Large sized dual purpose sire breeds behave like the continental beef sire breeds as regards growth potential. However those in which the double muscle trait has not been selected produce crossbred young bulls with a little lower carcass value. It is important to emphasise that the differences observed between all these sire breeds partly depends on the sample of sires of each breed as well as on present selection trends in each breed. These may perhaps explain the small difference in this comparison between sires of the T.C. type and those of P.B. type.

The interest of each of these breeds or genetic types of sire breeds for industrial crossing is closely related to the system of valorisation of the crossbred products (males and females) and of their respective potentialities. The rationalisation of the choice of sire breeds according to the beef production system from more (FR breed) or less (dual purpose breeds) specialised dairy herds should be studied more thoroughly. It is owing to such studies of industrial crossing systems induced by the results of comparisons between various sire breeds that the EEC may reach an optimum milk and meat production with its dairy herd, and may rapidly adapt itself to changes in their production systems.

References

Anderson, J., Lindhe, B. 1973. Optimum use of beef semen in a dual purpose or dairy breed. Acta Agric. Scand. 23, 102-108.

Belic, M., Menissier, F. 1968. Etude de quelques facteurs influençant les difficultés de vêlage en croisement industriel. Ann. Zootech. 17, 107-142.

Bibe, B., Frebling, J., Menissier, F. 1977. Double muscle sires for terminal crossing : French experiments. In : "Crossbreeding experiments and strategy of beef utilisation to increase beef production," EEC Seminar, Feb. 1976, Verden, FRG, (EUR. 5492e), 72-98.

Bonaiti, B. 1980. Influence du type génétique sur les caractéristiques et l'utilisation technologique de la carcasse de jeune bovin. Final report of D.G.R.S.T. contract (No. 78.7. 0467 and 0.468), 14p. (not published).

Cunningham, E.P. 1974. The economic consequences of beef crossing in dual purpose or dairy cattle populations. Livestock Prod. Sci. 1, 133-139.

Cunningham, E.P., McClintock, A.E. 1974. Selection in dual purpose cattle populations : Effect of beef crossing and cow replacement. Ann. Genet. Sel. anim. 6, 227-239.

De Boer, H., Dumont, B.L., Pomeroy, R.W., Weniger, J.H. 1974. Manual on EAAP reference methods for the assessment of carcass characteristics in cattle. Livestock Prod. Sci. 1, 151-164.

Elsen, J.M., Mocquot, J.C. 1976. Optimisation du renouvellement des femelles dans les troupeaux laitiers soumis au croisement terminal. Ann. Génét. Sél. anim. 8, 343-356.

Foulley, J.L. 1976. Some considerations on selection criteria and optimization for terminal sire breeds. In "Optimization of cattle breeding schemes," EEC Seminar, Nov. 1975, Dublin, Ireland (EUR. 5490e), 85-104. Ann. Genet. Sel. anim. 8, 89-101.

Foulley, J.L., Gaillard, J., Menissier, F. 1978. Choix des taureaux de races à viande pour la production de veaux de boucherie. In : "Le veau de boucherie", Ed. INRA publication, Bull. Tech. CRVZ de Theix (Inst. Nat. Rech. agric., Fr.), (No. special), 19-30.

Foulley, J.L., Menissier, F., Gaillard, J., Nebreda, A.M. 1975. Aptitudes maternelles des races laitières, mixtes, rustiques et a viande pour la production de veaux de boucherie par croisement industriel. Livestock Prod. Sci. 2, 39-49.

Foulley, J.L., Menissier, F. 1981. Selection of beef bulls for terminal crossing. In : "Beef production from the dairy herd," EEC Seminar, April 1981, Dublin-IR.

Frebling, J., Gaillard, J., Vissac, B. 1972. Il-Mise en place et efficacite du schéma de sélection des taureaux de races à viande pour le croisement industriel. In : "Efficacite reelle et optimum du choix des taureaux de races à viande pour le croisement industriel." Bull.Tech. Dept. Génét. anim. (Inst. nat. Rech. agr. Fr.) (No.15), 23-55.

Frebling, J., Poujardieu,B., Vissac, B., Azan, M., Gaillard, J., Rondeau, M., Hennequin, M. 1970. Station de sélection bovine. Compte rendu No. 6 : Comparaison des races Charolaise et Blonde d'Aquitaine en croisement de première génération sur la race d'Aubrac. Bull. Tech. Inf. Ingrs. Serv. agric., 253, 635-642.

Gaillard, J., Foulley, J.L.,Menissier, F. 1974. Observations sur l'efficacité du choix sur ascendance paternelle et performances individuelles de taureaux de race à viande destinés au croisement terminal. 1er Congrès Mondial de Génétique appliquée a l'élevage, octobre 1974, Madrid, (Vol. III), 877-887.

Henderson, C.R. Jr., Henderson, C.R. 1979. Analysis of variance in mixed models with unequal subclass numbers. Commun. Statist. Theor. Meth. A8(8), 751-787.

Menissier, F. 1980. Advantage of using double muscled sires in crossbreeding and selection of a specialized double muscle sire lines in France. In: "Muscle hypertrophy of genetic origin and its use to improve beef production." EEC Seminar, June 1980, Toulouse-Fr. (in press).

Menissier, F., Foulley, J.L. 1979. Present situation of calving problems in the EEC.: Incidence of calving difficulties and early calf mortality in beef breeds. In: "Calving problems and early viability of the calf." EEC Seminar, May 1977, Freising, FRG. Current Topics in Veterinary Medicine and Animal Science, 4, 30-85.

Menissier, F., Vissac, B., Frebling, J. 1974. Optimum production plans for milk and meat : specialised beef herds and beef crossing in dairy herds : 25eme Reunion Annuelle, Federation Europeenne de Zootechnie, Copenhague, Aout 1974. In : "Optimum breeding plans for beef cattle," Bull. Tech. Dept. Genet. anim. (Inst. nat. Rech. agric. Fr.) No.21,2-56.

Mocquot, J.C. 1972. III-Recherches sur la rentabilité optimale du schéma de sélection des mâles de races à viande pour une production de veaux et/ou de jeunes bovins. In: "Efficacité réelle et optimum du choix des taureaux de races à viande pour le croisement industriel." Bull. Tech. Dept. Génét. anim. (Inst. nat. Rech. agr. Fr.) No. 15,56-76.

Mocquot, J.C., Foulley, J.L. 1973. Recherche des conditions de rentabilité d'un schéma de sélection d'une souche de bovins destinee au croisement de première génération pour la production de veau de boucherie. Ann. Génét. Sél. anim. 5, 189-209.

Philipsson, J. 1977. Studies on calving difficulty, stillbirth and associated factors in Swedish cattle breeds : VI-Effects of crossbreeding. Act. Agric. Scand. 27, 58-64.

Poujardieu, B., Vissac, B. 1968. Etude biométrique de la valeur bouchère de veaux croisés Charolais et Limousins. I-Paramètres génétiques et phenotypiques. Ann. Zootech. 17, 143-158.

Robelin, J., Geay, Y. 1975. Estimation de la composition de la carcasse des taurillons à partir de la composition de la 6ème côte. Bull. tech. du C.R.V.S. de Theix (Inst. nat. Rech. Agr. Fr.). No. 22, 41-44.

Schaeffer, L. 1979. Estimation of variance and covariance components for average dairy gain and backfat thickness in swine. In Van Vleck, L.D. and Searle, S.R. "Variance components in animal breeding", Cornell Univ. Ithaca, New York, 123-137.

Tayler, J.C. 1976. Beef production in the EEC and the co-ordination of research by the Com mission of the European Communities. Livestock Prod. Sci. 3, 305-318.

Thompson, R. 1979. Sire evaluation. Biometrics, 35, 339-353.

Vissac, B. 1972. I-Sélection des souches mâles de bovins à viande pour le croisement terminal. In: "Efficacité réelle et optimum du choix des taureaux de races à viande pour le croisement industriel." Bull. Tech. Dept. Génét. anim. (Inst. nat. Rech. agr. Fr.), No. 15, 1-22.

Vissac, B., Bonhomme, D., Frebling, J. 1971. L'utilisation des races à viande francaises en croisement de première génération pour la production de veaux de boucherie : Bilan de dix années de recherches (1960-1970). Bull. Tech. Dept. Génét. anim. (Inst. nat. Rech. agr. Fr.), No. 12, 27p.

Vissac, B., Frebling, J., Faucon, A. 1965. Statistiques générales sur la production de veaux de boucherie en croisement industriel dans le Centre et le Sud Ouest de la France. Bull. Tech. Ingrs. Servs. agric., 204, 889-939.

Vissac, B., Poly, J., Charlet, P. 1959. Les épreuves de descendance de taureaux d'insémination sur la valeur de leur veaux de boucherie. Bull. Tech. Inf. Ingr. Servs. Agric. 145, 759-787.

BEEF PRODUCTION IN DANISH JERSEY : CHAROLAIS x JERSEY CROSSINGS

A. Neimann-Sørensen and E.O. Neilsen

National Institute of Animal Science, Rolighedsvej 25,
1958 Copenhagen V, Denmark

Summary

From a limited importation of a few thousand animals from the island of Jersey around the year 1900, the Danish Jersey breed has expanded to 150,000 cows. In respect of butterfat production, the Jersey out-yields all other breeds, and its small size, good constitution and high longevity make it one of the most efficient dairy breeds of to-day.

Its obvious lack of size and muscularity limits its value in beef production, but extensive experimental data as well as the practical experience of many farmers over the last twenty years show the advantages in the latter respect which can be obtained by the use of Charolais bulls/semen in a commercial crossbreeding programme. When heifers are excluded, neither the calving performance nor the milk production parameters will be negatively influenced to any appreciable degree, and the primary outcome of the crossing - the offspring for slaughter - will show growth, carcass and meat characteristics which will compare favourably with those of the large (dual-purpose) dairy breeds of to-day.

Introduction

The first Jersey animals came to Denmark in 1896. From then and up to 1909, altogether 5,303 animals were imported from the island of Jersey. Since then a few bulls were imported from Jersey in 1948-51 (6 bulls) and again in 1970-71 (3 bulls), but none of these produced progeny of a satisfactory yield, and their impact on the breed was negligible.

Since 1960, semen has been imported from a few selected bulls of high genetic merit from New Zealand and the USA, and in 1980 semen was also imported from Canada. Semen from these importations was mainly used with some of the best cows within the breed, and as an outcome of this, a few meritorious bulls have been produced and used in AI.

From the above very limited foundation stock, which people from the island of Jersey named as "rubbish", the present breed of Danish Jersey has been developed. The breed at present totals 175,000 milk cows, plus young stock, yielding 4,100 kg milk, 6.25% fat and 256 kg butterfat (Table 1). The breed numbers have increased nearly twenty times in the years from 1945 to 1960, and this explosive expansion of the breed is unique. It illustrates that a cattle breed which possesses a good constitution and longevity, high yielding capacity, supported by a determined and well-defined breeding goal of its breeders, is able to expand and at the same time maintain a continuous rise in yield. During later years the Jerseys have practically kept their proportion of the total cattle population.

As a result of the 80 years of isolated breeding, the Danish Jerseys differ from Jerseys in other countries mainly in two characters, viz. they are smaller in size and weight, and produce milk of a higher fat percentage. Table 1 highlights some of the characteristics of the Danish Jersey breed of to-day, compared with the Red Danish (RDM) and the Black and White (SDM) breeds.

It is seen from Table 1 that the Jersey out-yields the two large breeds in butterfat production, as it out-yields any other breed in the world in this respect. This capacity, combined with the low cost of raising heifers, the low maintenance requirement and high fat percentage of the milk, all contribute to the high efficiency in butterfat production of the breed. At the same time, it may be noted that the Jersey cow shows favourable figures regarding milking ability, ease of calving and incidence of disease, altogether leading to a remarkable record in lifetime production of butterfat.

Calving difficulty is more rare in Jersey cows than in cows of the heavy dairy breeds, as is apparent from Table 2. This table gives information about the calving performance of the bull progeny test stations in 1974-75, the last year the stations were in use. The results refer to first-calving cows. In subsequent calvings, difficulties were rare and of little practical importance.

Liboriussen (1978) found in a crossbreeding experiment with sires from eight different European beef breeds used on cows (different parities) of the breeds RDM, SDM and Jersey, that

Table 1 Average values of different traits in Danish cattle breeds

	Breed		
	Red Danish	Black and White	Jersey
Milk yield in 305 days, kg	5500	5600	4200
Butterfat in 305 days, kg	231	226	265
Fat percentage	4.2	4.0	6.3
Protein percentage (1. lact)	3.8	3.7	4.3
Mature weight of herdbook cows, kg	675	700	410
Weight of bulls, 1-year-old, kg	460	480	300
Age at first calving, months	29	29	26
Feed units/heifer raised to calving	3100	3100	2400
Feed units/kg butterfat produced	17	17	15
Milking time, min	5.2	4.9	4.1
Percentage of milk in forequarters	45	44	47
Incidence of mastitis, per 100 cows/year	85	47	43
Incidence of milk fever " " " "	12	8	18
Incidence of ret. placenta " " "	15	13	5
Incidence of total diseases " " "	194	138	108
Cows with total yield > 3500 kg bf frequency per 100 cows	0.04	0.04	1.00
Generation interval. Cow-cow, years	5	5-6	6-7

Table 2 Calving registrations at Danish bull progeny test stations 1974-75. (Nielsen et al. 1975)

Breed	No. cows	No. calves		No. mis-placed calves	No. calvings			Vet. assist-ance	Dissection or caesarian
		Born alive	Still-born		nor-mal	slight-ly dif-ficult	Very dif-ficult		
RDM	249	204	45	23	204	23	22	24	7
SDM	280	260	20	22	234	26	20	20	3
Jersey	154	146	8	3	150	4	0	0	0

the Jersey cows exhibited a significantly higher proportion of easy calvings than the heavy breeds (Table 3). In the experiment, the calving performance was classified in the following way :

1 = easy calving. No assistance needed.

2 = difficult calving. Assistance needed but no veterinarian present.

3 = very difficult calving. Veterinary assistance needed.

Table 3 Average figures for gestation length, birth weight and calving performance when sires of eight different beef breeds were used in crossbreeding with Red Danish (RDM), Black and White Danish (SDM) and Jersey cows. (Liboriussen 1978)

Breed of dam	No. animals	Gestation length	No. animals	Birth weight	No. animals	Average score for ease of calving
RDM	429	286.5	429	46.3	454	1.60
SDM	393	284.5	393	44.6	420	1.64
Jersey	123	284.7	123	32.6	130	1.25

It is apparent that the Jersey cows show less difficulties in calving than the other two breeds. This is remarkable, as the "overweight" of the crossbreed calves represents 16%, 12% and 25% respectively for the three breeds RDM, SDM and Jersey. Obviously, there could be serveral reasons for the superior calvability of Jerseys as compared to other breeds. One factor seems to be the anatomy of the pelvis. Nielsen (1965) studied this in detail and found the results shown in Table 4.

It is seen that relative to weight and height of the cows, the two large breeds do not differ significantly in their internal skaletal pelvic measurements. In contrast, these measurements are proportionally larger in the case of the Jersey cow, indicating a more roomy pelvis, and this may well be one of the major components for the superior calving performance ability of this breed. Nielsen (1965) also found

Table 4 Internal pelvic skeletal measurements of cows of the Red Danish (RDM), Black and White Danish (SDM) and Jersey breeds (Nielsen 1965)

	Young cows			Mature cows		
	RDM	SDM	Jersey	RDM	SDM	Jersey
Weight, kg	501	512	307	557	579	383
Height at withers, cm	128	128	118	129	129	120
Internal skeletal pelvic measurements						
Height, conj. vera. cm	22.6	21.5	21.5	24.2	23.4	23.3
% of weight	4.5	4.2	7.0	4.3	4.0	6.1
% of wither height	17.7	16.8	18.2	18.8	18.1	19.4
Width, max. cm	18.5	18.8	16.3	19.5	19.9	17.1
% of weight	3.7	3.7	5.3	3.5	3.4	4.5
% of wither height	14.5	14.6	13.8	15.1	15.4	14.3

significant correlations between the internal skeletal measurements and the hindquarter conformation of cows (breadth and flatness of the rump), and this points to the possibility that the conformation ideals of the past may have been purely aesthetically, rather than functionally founded.

Beef production from Jerseys

It is a matter of self evidence that the capacity of beef production from the Jersey is much less than that of the large dairy breeds (Table 1). Growth rate, muscularity and final weight is inferior by 25-30%. On the other hand, early breeding capacity, longevity and low disease incidence in the Jersey offer this breed good opportunities for exploiting the possibilities of using commercial crossbreeding with beef breeds. In Denmark the culling percentage (voluntary and involuntary) goes up to around 40%, whereas it seldom exceeds 30% in the Jersey herd. This implies that hardly any crossbreeding is possible in the large dairy breeds, whereas potentially 30-35% of the Jersey cows could be used for crossbreeding with beef bulls without interfering with the maintenance of herd size.

The Jersey breeders were early aware of this, and over the

years many experiments have been carried out investigate the possibilities and to identify the most suitable breed(s) to cross with Jersey cows in order to improve beef production from this breed. In 1962 an experiment was carried out using Aberdeen Angus, Hereford, Red Danish and Charolais as sire breeds. Both male and female offspring were included. Indoor feeding with a basic ration of beet and concentrates was applied. The animals were put on experiment at 45 days of age and two slaughter weights were used, viz. approximately 200 kg and 350-400 kg. The results are presented in Tables 5 and 6.

Table 5 Four sire breeds used in crossbreeding with Jerseys. Results from crossbred animals slaughtered at approximately 200 kg liveweight (Larsen et al. 1962)

	Sire breed							
	AA		Hereford		R.Danish		Charolais	
	♂	♀	♂	♀	♂	♀	♂	♀
No. of animals	6	8	7	8	11	7	9	10
Initial wt kg	50	50	51	50	58	51	50	50
Wt at slaughter kg	199	195	205	206	221	205	218	212
Daily gain g	811±25	709±11	906±20	775±11	971±17	762±15	981±17	804±1
Sc. f.u. per kg gain	3.72	4.23	3.88	4.40	3.86	4.52	3.54	4.0
Dressing percentage	53.8	54.0	53.2	53.3	54.4	44.0	56.5	58.2
Carcass eval. points[1]	8.0	8.5	8.4	7.9	7.5	7.0	8.0	7.1
Meat in carcass, %	74.3	72.3	72.7	70.0	73.6	71.0	74.8	72.5
Fat in carcass, %	8.6	11.7	9.3	13.3	7.9	12.2	7.8	11.1
Bone in carcass, %	17.1	16.0	18.0	16.7	18.5	16.8	17.4	16.4
'Pistol cut'[2] %	50.2	49.4	48.7	48.5	47.9	47.2	47.6	48.1

[1] Scale 1-10

[2] Back (from the 6th costal vertebra) + thigh

Table 6 Four sire breeds used in crossbreeding with Jerseys. Results from crossbred animals slaughtered at 350-400 kg liveweight (Larsen et al. 1962)

	Sire breed							
	AA		Hereford		R.Danish		Charolais	
	♂	♀	♂	♀	♂	♀	♂	♀
No. of animals	9	9	8	8	11	6	8	10
Initial wt kg	50	50	52	52	54	51	48	48
Wt at slaughter kg	385	311	391	317	414	321	396	323
Daily gain g	858±23	702±23	915±20	771±13	955±26	774±19	1007±20	808±13
Sc.f.u. per kg gain	4.74	5.41	4.80	5.04	5.18	5.23	4.59	4.92
Dressing percentage	57.2	56.3	54.4	56.8	56.1	56.4	59.0	57.9
Carcass eval. points[1]	8.0	7.4	8.0	7.5	6.7	6.8	8.0	7.2
Meat in carcass, %	74.6	68.3	71.0	64.0	73.0	65.9	73.9	68.1
Fat in carcass, %	10.8	17.5	13.5	19.3	11.0	18.6	11.4	17.6
Bone in carcass, %	14.6	14.2	15.5	15.2	16.0	15.5	14.7	14.3
'Pistol cut', %	44.7	46.8	45.6	46.3	45.4	46.0	45.9	47.2

[1] Scale 1-10

[2] Back (from the 6th costal vertebra) + thigh

The tables are an illustration of the effects of sex, weight at slaughter (age) and sire breed in Jersey crossings. It will be seen that the two British breeds, Aberdeen Angus and Hereford, are the least suitable. In the case of the Aberdeen Angus, this is mainly due to an unsatisfactory growth capacity which is most pronounced at the higher slaughter weight. The Hereford crossbreds show a little higher growth capacity than the Aberdeen Angus crosses but they show a pronounced aptitude for fat deposition which at higher weights can be quite excessive. Both at the low and the higher weight at slaughter do the Charolais crossbreds exhibit the best results. The growth rate both in male and female calves is significantly higher than for any of the other crosses, and it is in nearly all cases superior in

Table 7 A summary of the beef production results, including Jerseys and Jersey x Charolais bulls in the Danish performance and progeny tests of young bulls for AI

Performance tests

Breed	No. animals	Initial wt. (kg) at 42 days	Final wt. (kg) at 336 days	Daily gain (g)	Feed conversion SFU/kg gain[1]	Height of withers (cm)
Jersey	241	38.6	285	836	4.14	111
Red Danish	833	62.3	409	1179	4.42	121
Black & White Danish	195	64.5	430	1243	4.22	122

Progeny tests

Breed	No. animals	Wt. at slaughter (kg)	Daily gain (g)	Feed conversion SFU/kg gain[1]	Dressing %	Grading[2] score	% kidney fat	% fat in meat	Lean/bone[4] in pistol cut	Tenderness[3] of meat
Jersey	28	284	1027	3.8	50.6	4.2	1.5	2.2	4.1	7.2
Charolais x Jersey	113	324	1221	3.5	52.9	6.8	1.1	1.4	4.5	8.7
Red Danish	542	324	1266	3.6	52.2	5.4	1.0	1.5	3.8	7.9
Black & White	359	324	1293	3.5	52.4	7.1	0.9	1.2	3.8	9.3

1 SFU = Scandinavian feed unit

2 Scale 1 - 10

3 Shear force

4 Different number of animals, viz. 28, 36, 188 and 108, resepectively

carcass characteristics. The results from the Red Danish crosses come close to those of the Charolais crosses.

It should be noted that the females in all cases show lower (approx. 20%) growth rate, lower feed efficiency and enhanced inclination for fat deposition than the male calves. In a commercial crossbreeding programme this must be taken into consideration, as it inevitably leads to a lower economic outcome than that obtainable from fattening of male calves.

The results presented, together with similar observations of many farmers in practice, made the Danish Jersey breeders almost exclusively choose Charolais bulls whenever crossbreeding was practised. As a further consequence of this, several Charolais bulls have been progeny tested on crossbred bull calves in recent years. Table 7 is a compilation of the results obtained in these tests, and for comparative results from pure Jersey and from the heavy dairy breeds Red Danish and Black and White Danish are presented in the same table.

Without going into details about the results presented in Table 7, the following major inferences can be drawn. Compared to the large dairy breeds, the purebred Jersey calves show a growth rate of 20-30% less, their dressing-out percentage and grading score are somewhat inferior, but nevertheless, the lean : bone ratio is higher and the eating quality of their meat is more favourable, as expressed by the higher fat content and tenderness of the meat. Again, compared to the large dairy breeds, it is remarkable that the Charolais crossbred calves perform so well. They are equal in growth rate, feed conversion and dressing-out percentage, but clearly superior in carcass conformation (grading) and lean : bone ratio.

In the light of the above results, the Danish Jersey breeders have practised commercial crossbreeding of their cows with Charolais bulls to a considerable extent over the years. This practice was at its height in the late Sixties and the beginning of the Seventies, when milk prices were low compared to meat prices, as Denmark was at that time outside the EEC and thus dependent on world market prices for its export outlets. It was not uncommon to see individual herds where 20-30% of the cows were bred to beef bulls, mainly Charolais, cf. Figure 1.

FIGURE 1 The extent (in percent) of commercial crossbreeding with Charolais bulls in Danish Jersey in the years 1971-1980

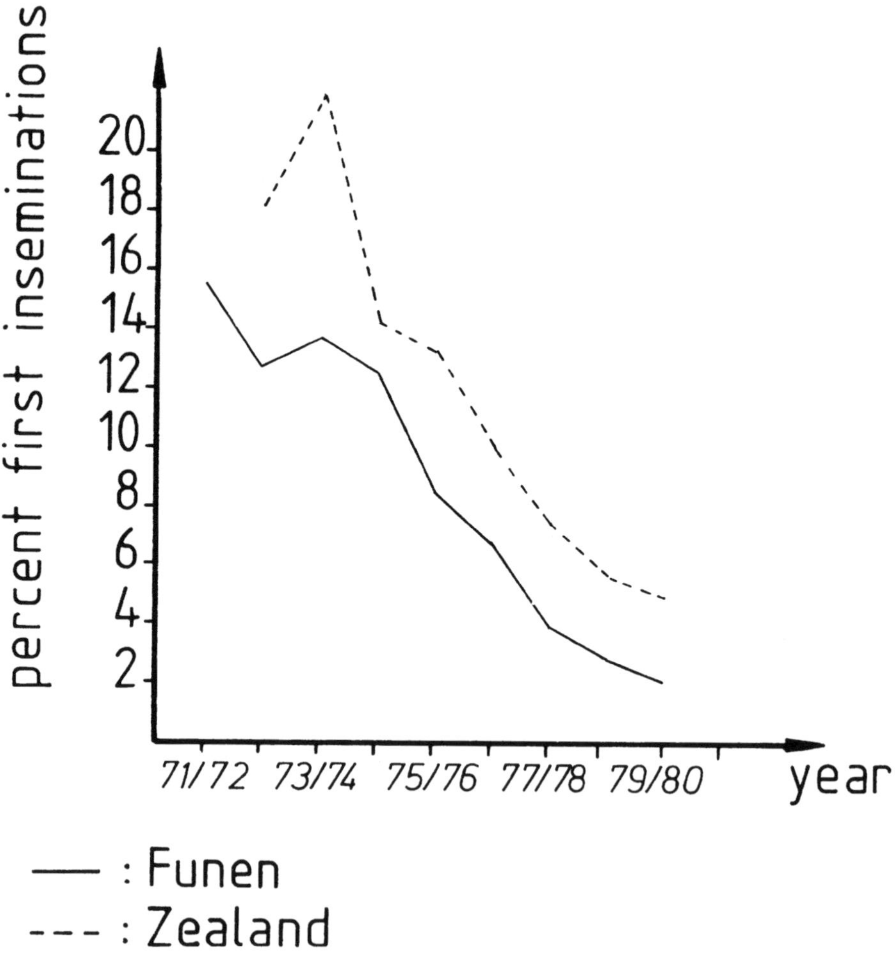

As can be seen from this Figure, there has been a dramatic decline in crossbreeding since 1972/73 and at present it does not exceed 5% of the Jersey cows, taken as an average. This development must be taken as a natural reaction to the rise in milk prices compared to the meat prices in the EEC in this period. Undoubtedly, however, other factors have also played a role. Several farmers, especially the owners of smaller herds, found that the practice of crossbreeding close to the biologically determined limit brought them, due to chance, into situations where the herd could not maintain itself; the demand, for domestic and export purposes, for Jersey heifers of high genetic merit has been another factor. Finally, several farmers, Jersey breeders or others, have realised that bull calves of this breed, when properly fed and slaughtered at optimum weight, can still form the basis for an economically and qualitatively sound beef production.

The effect of Charolais crossbreeding on milk production in Jerseys

Over the years it has repeatedly been claimed by Jersey farmers practising crossbreeding with Charolais bulls, that cows pregnant to a Charolais bull had lower yields and became dry earlier than cows pregnant to Jersey bulls. In 1972 Forsyth and Buttle were able to show that a hormone, placental lactogen, produced in the foetal part of the placenta, was active in ruminants. This hormone was later isolated from cows and characterised by Bolander and Fellows (1976) and Buttle and Forsyth (1976). Since then, several studies have indicated that this hormone may represent a way in which the genotype of the foetus may influence the yield of cow, both in the simultaneous and the subsequent lactation (Skjervold & Finland 1975; Adkinson *et al*. 1977; Auran *et al*. 1977; Johnson & van Vleck, 1978). In all cases the effects were small - of the order of 5-10% of the total variance in yield and less than that in the variance of the length of the dry period.

In the years 1974-78 an experiment with the aim of elucidating these possibilities in the case of Jerseys crossed with Charolais was carried out in Denmark (Nielsen & Jorgensen, 1979) including two groups of paternal half-sisters of first lactation Jersey cows. The plan of the experiment (split-pair) was as follows :-

1. lactation		2. lactation		3. lactation	
Exp. group No.	Sire of foetus	Exp. group No.	Sire of foetus	Exp. group No.	Sire of foetus
1	Jersey	3	Jersey	7	Jersey
		4	Charolais	8	Jersey
2	Charolais	5	Jersey	9	Jersey
		6	Charolais	10	Jersey

As will be seen, half of the cows in the first lactation were inseminated with semen from a Jersey bull, while semen from a Charolais bull was used to the remainder cows. In the second lactation, the groups were divided once more, while all cows were inseminated by semen from Jersey bulls in the third lactation. The results are summarised in Table 8.

Table 8 The effect of using Charolais bulls on Jersey cows on milk production in the current and subsequent lactation (summarised results from Nielsen & Jorgensen 1979)

The effect on the current lactation (first and second) :

Pregnant with	Exp. group Nos.	No. lactations	Days open	Days in milk	305 days yield Butterfat kg	Days dry
Jersey	1 + 3 + 5	34	79	307	274	54
Charolais	2 + 4 + 6	35	109	334	274	60

The effect on the subsequent lactation (second and third) :

Pregnant with	Exp. group Nos.	No. lactations	Days open	Days in milk	305 days yield Butterfat kg	Days dry
Jersey	3+4+7+9	29	88	317	302	56
Charolais	5+6+8+10	28	79	297	288	65

Concerning the effect on the current lactation the average butterfat yield was 274 kg for both groups, but the cows pregnant with Charolais had on average 30 more days open than those pregnant with Jersey, due to the late pregnancy of some cows. As to the influence of the foetus on the subsequent lactation, a difference of 14 kg butterfat was found between the two groups.

For none of the milk production parameters, days in milk, 305 days butterfat yield, or days dry, do the results presented in Table 8 represent differences which are statistically significant. It can be concluded that any effect which the use of Charolais bulls on Jersey cows may have on milk production is small, although it cannot be excluded that the effect may be more pronounced under practical circumstances.

References

Auran, T., Finland, E. and Skjervold, E. 1977. The fetal effect on milk yield in current lactation. E.A.A.P. 28th Ann. meeting, pp. 8.

Bolander, F.F. and Fellows, R.E. 1976. Growth promoting and lactagenic activities of bovine placental lactogen. J. Endocrinology, 71, 173-174.

Buttle, H.L. and Forsyth, I.A. 1976. Placental lactogen in the sow. J. Endocrinology, 68, 141-146.

Forsyth, I.A. and Buttle, H.L. 1972. Placental lactogen in the ruminant. IV Int. Congr. Endocrin. (abstr.). Excerpta Med. Wash. D.C. p. 106.

Johnson, L.P. and van Vleck, L.D. 1978. Components of variance for effects associated with sire of fetus and service sire on milk yield, gestation length and days open (abstr.) J. Dairy Sci. 61, suppl. 1, 87.

Larsen, J.B., Klausen, S. and Lykkeaa, J. 1961. Krydsningsforsøg med Jersey. Arbog fra Landøkonomisk Forsøgslaboratorium. p. 52-55.

Liboriussen, T. 1978. Brugskrydsning i SDM og RDM. 466. Beretning fra Statens Husdyrbrugsforsøg. 123 pp.

Nielsen, E. and Jørgensen, J. 1979. Jerseykøer drægtige ved tyre af racerne Jersey og Charolais. 481. beretning fra Statens Husdrybrugsforsøg. 40 pp.

Nielsen, E. and Vesth, B. 1975. Afkomsprøver med tyre XXIX. 423. Beretning fra Forsøgslaboratoriet. 148 pp.

Nielsen, J. 1965. En undersøgelse over sammenhængen mellem krydsets og bækkenindgangens dimension hos køer. Årbog fra Landøkonomisk Forsøgslaboratorium. p. 239-260.

Skjervold, H. and Fimland, E. 1975. Evidence for a possible influence of the fetus on the milk yield of the dam. Z. Tierzucht. u. Zucht. biol. 92, 245-251.

DISCUSSION ON DR. A. NEIMANN-SORENSEN'S PAPER

Chairman: Dr. R. Thiessen

Cunningham: I would like to add a word on an experiment which we have recently completed and which confirms the results presented by Professor Neimann-Sorensen. A total of approximately 2000 cows in 19 Friesian herds were mated at random to Friesian, Charolais, Hereford, Angus, Simmental and Limousin bulls. The overall results indicated no significant effect of the breed of bull on either the current lactation or the subsequent lactation.

Jansen: Can you explain why differences were so large for the number of days open ?

Neimann-Sorensen: This was just an unfortunate result, arising through difficulties in getting some of the cows in calf. We have corrected the yield data for days open, but even so, the observed yield differences were not statistically significant.

Oostendorp: Might there not also be situations in which a positive influence of calf on milk production could be observed ?

Neimann-Sorensen: Yes, for example in the Norwegian and United States studies which I have referenced in my paper, the studies on population data have shown that 5 - 10% of the within-breed variance in yield, and less than that for the variance of the length of dry period, can be attributed to the genotype of the foetus.

Zervas: How do farmers overcome replacement problems with small cow numbers, when crossing with beef bulls ?

Neimann-Sorensen: The greater longevity of Jerseys assists in this respect. The culling rate is only 25 - 30% and this implies that up to 30% of cows can be bred to beef bulls without interfering with the heifer replacement.

PERFORMANCE OF BELGIAN WHITE-RED CROSSBREDS USED FOR BEEF PRODUCTION - PRELIMINARY RESULTS*

L.O. Fiems, Ch.V. Boucqué and F.X. Buysse

Scheldeweg 68, B-9231 Melle-Gontrode, Belgium.

Summary

Fattening trials were carried out with Charolais, Piedmont, White-Blue and Beefalo crossbreds, born of Belgian White-Red females.

Daily liveweight gain during the finishing period amounted to 1025, 1323, 1282 and 1277 g respectively for bulls of the aforementioned crosses; growth rate was not increased compared with the purebred White-Red bulls. Charolais, Piedmont and White-Blue female crossbreds gained 612, 496 and 553 g daily compared with 578 g for the purebred White-Red heifers.

Crossbreeding with these sires resulted in a favourable feed conversion with the male offspring.

There was also a tendency for a higher dressing percentage with these crossbreds; the lower fat content in their carcass was in accordance with the better energy efficiency.

Introduction

In Belgium, five breeds are officially recognised, namely: the White-Blue, the White-Red, the Red Pied, the Red, and the Black and White breed. This means that financial support is at the disposal of the Herdbook Associations of these breeds for breeding and selection purposes. Other breeds like the Charolais, the Limousin, the Blonde d'Aquitaine, the Holstein Friesian and the Jersey, are admitted, but support is not available. Perhaps that is a reason for the low number of crossbreds in Belgium. Another factor explaining the low number of crossbreds might be the presence of the beef strain of the Belgian White-Blue breed, with its good beef production ability due to the double-muscled animals (Hanset, 1980). This, indeed, is the most important breed.

* Communication No. 459, Institute of Animal Nutrition.

Nevertheless, in dairy herds more heifer calves are born than necessary for a normal cow replacement rate. With three lactations per cow and 0.8 heifers per cow destined for culled cow replacement, a maximum of 17% of the dairy cows can be used for crossbreeding without decreasing the dairy herd (Cunningham, 1974; Politiek, 1975).

The importance of the different breeds in Belgium is given in Table 1. The number of crossbred cows used for beef production amounts only to 0.9% of the cow population.

Table 1 Number of cows per breed in Belgium

Breeds		Number	%	
Beef breeds				
- Charolais		9,376	0.8	
- Limousin		2,655	0.2	
- Blonde d'Aquitaine		868	0.1	
- Belgian White-Blue* (beef strain)		109,851	9.9	
- Crossbreds		9,918	0.9	
	132,668			11.9
Dual-purpose breeds				
- Belgian White-Blue* (dual-purpose strain)		288,160	25.9	
- Belgian White-Red*		109,958	9.9	
- Belgian Red*		70,875	6.3	
	468,993			42.1
Dairy breeds				
- Belgian Red-Pied* (MRY)		235,660	21.2	
- Belgian Black and White*		248,794	22.3	
- Holstein Friesian		5,321	0.5	
- Jersey		507	<0.1	
- Crossbreds		13,799	1.2	
	504,081			45.3
Other breeds		7,529	0.7	
Total number of cows		1,113,271	100	

* Officially recognised Source: Census 15/5/1980 (N.I.S.)

2 Experimental design

Fattening experiments were carried out with the following animals :- Charolais, White-Blue, Piedmont, and Beefalo crossbreds. They were born of Belgian White-Red primiparous females, except the Beefalo crossbreds. Beefalo sires were a combination of 3/8 buffalo, 3/8 Charolais and 1/4 Hereford. The last mentioned animals passed a rearing period of about six months based on an artificial milk replacer. The other crossbreds suckled their dam during the first six months of life. Afterwards the bulls were intensively fattened on a dry ration, based mainly on dried sugar beet pulp. The diet was freely available to obtain better information about their genetic ability for growth, feed intake and efficiency. The performance of the crossbreds was compared with that of the purebred White-Red bulls, fattened in identical circumstances after a six-month suckling period. All the bulls were group-housed in a straw bedded pen.

After weaning, the heifer calves were extensively fattened. They were turned out on to pasture during the summer period and were supplemented with dried sugar beet pulp during their early lifetime and when there was a lack of grass. The ration during the indoor period consisted of herbage products and maize silage.

All animals were slaughtered when they had a proper degree of fatness. Carcass composition was assessed by an 8th rib cut (Verbeke & Van de Voorde, 1978).

3 Results and Discussion

3.1 Animal performance

In the first comparative fattening trial the Charolais crossbreds and the purebred White-Red bulls received a diet *ad libitum* consisting of half-dried sugar beet pulp and half alfalfa pellets, supplemented with 1 kg concentrate daily.

Their performances are summarised in Table 2. During the suckling period both groups had a similar liveweight gain, although the White-Red calves were not suckled by their own mothers, but they substituted stillborn calves. This could suppose that there were not problems of adaptation, since this can delay daily gain (Le Neindre & Petit, 1976) but the

Table 2 Performance of crossbred bulls ($\pm s_{\bar{x}}$)

	White-Red	Charolais	White-Red	Piedmont	White-Blue	Beefalo
Number	6	12	11	6	4	5
Suckling period						
Birth weight (kg)	52.0	47.3	46.5	44.5	44.5	43.6
Daily gain (g)	$1205^{a} \pm 141$	$1199^{a} \pm 36$	$1106^{ab} \pm 52$	$1203^{a} \pm 24$	$879^{bc} \pm 159$	$774^{c} \pm 47$
Fattening period						
Initial weight (kg)	254.2	247.8	236.0	251.0	202.0	173.4
Final weight (kg)	633.8	606.8	596.8	615.5	636.0	543.6
Days	320.8	350.3	270.1	275.5	338.5	290.0
Daily gain (g)	$1183^{a} \pm 68$	$1025^{b} \pm 41$	$1336^{a} \pm 53$	$1323^{a} \pm 53$	$1282^{a} \pm 56$	$1277^{a} \pm 43$
Daily feed intake (kg)	10.17	8.30	8.88	8.53	7.87	7.99
Daily intake/kg $W^{0.75}$ (g)						
- dry matter	92.0	76.9	84.5	78.8	74.7	84.9
- starch units	53.6	46.1	64.1	59.8	56.7	64.5
Feed conversion (kg)						
- dry matter	7.50	7.06	5.83	5.66	5.40	5.48
- starch units	4.37	4.23	4.42	4.29	4.10	4.16
Birth to slaughter						
Mean daily gain (g)	$1191^{a} \pm 67$	$1081^{a} \pm 35$	$1247^{a} \pm 41$	$1277^{a} \pm 31$	$1142^{ac} \pm 61$	$1093^{bc} \pm 28$
Daily carcass gain (g)	718	642	795	808	723	632

a ... means on the same line with different superscripts are significantly different ($P < 0.05$)

variation in this group was larger. Afterwards the growth rate of the Charolais crossbreds was significantly lower ($P < 0.05$). This phenomenon was not expected since purebred Charolais bulls obtained a high daily gain during the finishing period (Geay & Béranger, 1975; Boucqué _et al._, 1977). Maybe daily gain was depressed due to the low energy density (± 590 SU/DM) as also stated by Geay and Beranger (1975). Positive results of Charolais used in crossbreeding were mentioned by Kilkenny (1975), Bech Andersen _et al._ (1977), and Tilsch _et al._ (1977).

Since the Charolais crossbreds were born of heifers calving at two years, this may have decreased their growth rate compared with the offspring of multiparous cows (Kilkenny, 1973).

The lower daily gain of the crossbreds can be explained by an inferior feed intake of 14%. Since Charolais bulls had a relatively low appetite (Geay & Béranger, 1975), and also the Charolais crossbreds in this trial, a ration with a high energy concentration seems absolutely necessary to obtain a high daily gain. After all, the feed conversion was still better for the crossbreds : 4.23 kg starch units against 4.41 kg for the White-Red bulls.

In a second fattening experiment, crossbreds of White-Blue, Piedmont and Beefalo sires were again compared with White-Red bulls. The complete dry ration contained 70% dried sugar beet pulp. During the suckling period the fastest liveweight gain was obtained with the Piedmont crossbreds. The lower gain of the White-Blue bulls was a consequence of sanitary problems (pneumonia) and a lower milk production of one dam. The daily gain of the Beefalo bull calves is quite normal since they were reared with a restricted amount of milk substitute (Boucqué _et al._, 1978). All of these groups realised a high daily gain during the finishing period, varying from 1,277 to 1,336 g. The differences between the groups were not significant ($P < 0.05$). Similarly, high rates of gain were also reported by Boucqué _et al._ (1976).

As in the first experiment, the White-Red bulls had the highest feed conversion once again : 4.42 kg starch units per kg gain. The best feed efficiency was obtained with the White-Blue crossbreds due to the fact that two bulls were double muscled animals. The high efficiency of the Beefalo crosses must be interpreted as a consequence of the lower weight interval. The higher daily gain observed in the second experiment is a consequence of the higher energy density in the diet : ± 760 g SU/DM against ± 590 for the first trial (Boucqué _et al._, 1980).

The fattening of the females indicated a higher liveweight gain of the Charolais crossbreds during the finishing period compared with the purebred White-Red heifers. The faster daily gain of the crossbreds during suckling can be explained from the effect of suckling as against artificial rearing based on a restricted amount of milk substitute. In spite of the better growth rate the crossbreds had an unfavourable feed conversion.

The White-Blue and the Piedmont crossbreds had a daily gain of 553 and 496 g respectively. This phenomenon is somewhat opposite to the liveweight gain of the male crossbreds. The feed conversion was better for the White-Blue crossbred females. This was in accordance with the results of the male crossbreds. The results are presented in Table 3.

3.2 Carcass data

Crossbreeding mostly resulted in an increase of the dressing percentage in comparison with the purebred White-Red animals, except for Charolais and Beefalo male crossbreds. The lower killing-out percentage of the Charolais male halfbreds was unexpected, based on the average percentage of more than 64% for Charolais purebreds (Geay & Béranger, 1975; Boucqué _et al._, 1977). The females, however, dressed-out better compared to the White-Red heifers. The dressing percentage of the Beefalos (59.5%) was not very interesting for Belgian circumstances, and was really lower than the 62% (based on the warm carcass weight !) mentioned in "Meat Industry" (Anon. 1975).

A better meat-fat relation was obtained with crossbreds, except again for the Beefalos. The best ratio was noted for the White-Blue halfbred bulls, due to the presence of double muscled animals, followed by the Piedmont crossbred bulls, while it was just the reverse with the females.

There was good agreement between the fat content in the carcass and the energy intake per kg liveweight gain. A less

Table 3 Performance of crossbred heifers

Sire	White-Red	Charolais	Piedmont	White-Blue
Number	9	6	4	8
Suckling period				
Birth weight (kg)	47.2	45.5	48.5	43.8
Daily gain (g)	$595^{a} \pm 39$	$1046^{b} \pm 56$	$949^{b} \pm 114$	$987^{b} \pm 49$
Fattening period				
Initial weight (kg)	130.6	222.7	215.5	213.4
Final weight (kg)	542.3	581.0	521.5	536.9
Days	712	585.2	616.8	585.5
Daily gain (g)	$578^{ac} \pm 17$	$612^{a} \pm 21$	$496^{b} \pm 37$	$553^{bc} \pm 12$
Daily intake				
- dry matter (*) (kg)	3.50	3.86	3.66	3.77
- starch units (*) (kg)	1.97	2.65	2.39	2.45
- pasture days	0.59	0.51	0.56	0.52
Feed conversion (kg)				
- dry matter (*)	6.05	6.30	7.37	6.83
- starch units (*)	3.40	4.32	4.81	4.44
Birth to slaughter				
Mean daily gain (g)	$581^{a} \pm 13$	$710^{b} \pm 23$	$597^{a} \pm 28$	$672^{b} \pm 19$
Daily carcass gain (g)	325	420	356	397

(*) nutrients from fresh grass (pasture period) are not included.

a, ... means on the same line with different superscripts are significantly different ($P < 0.05$)

favourable feed converion resulted in a fatter carcass, except for the Beefalo bulls and the Piedmont heifers. However, Beefalos had a lower weight interval and Piedmont heifers a lower daily gain than the other crossbreds.

The carcass data are summarised in Table 4. The positive effect of crossbreeding with Piedmont sires on the dressing percentage was also reported by Brunnekreef (1975), Fischer et

Table 4 Carcass characteristics of crossbreds

	White-Red	Charolais	White-Red	Piedmont	White-Blue	Beefalo
Bulls						
Dressing percentage	$62.3^{ab}\pm0.6$	$61.3^{ab}\pm0.9$	$61.2^{ac}\pm0.4$	$64.7^{b}\pm0.5$	$64.7^{bc}\pm2.6$	$59.5^{a}\pm2.2$
Carcass composition						
- % meat	$62.8^{a}\pm1.9$	$66.6^{c}\pm0.6$	$63.4^{a}\pm1.0$	$69.7^{bc}\pm1.2$	$71.4^{b}\pm1.5$	$60.6^{a}\pm1.2$
- % fat	$25.3^{a}\pm2.2$	$20.3^{c}\pm0.8$	$23.1^{ac}\pm1.1$	$16.4^{b}\pm1.1$	$14.7^{b}\pm1.4$	$26.0^{a}\pm0.9$
- % bone	$11.9^{a}\pm0.5$	$13.1^{ab}\pm0.4$	$13.5^{b}\pm0.4$	$13.9^{b}\pm0.4$	$13.9^{b}\pm1.2$	$13.4^{ab}\pm0.6$
Heifers						
Dressing percentage	$58.1^{a}\pm0.6$	$60.8^{ac}\pm1.3$		$60.9^{bc}\pm0.4$	$60.7^{bc}\pm0.4$	
Carcass composition						
- % meat	$60.5^{ac}\pm1.0$	$58.9^{a}\pm0.5$		$65.5^{b}\pm1.7$	$62.5^{bc}\pm0.8$	
- % fat	$26.7^{ac}\pm1.1$	$29.4^{a}\pm0.6$		$22.2^{b}\pm1.8$	$24.5^{bc}\pm1.1$	
- % bone	$12.8^{a}\pm0.3$	$11.7^{b}\pm0.2$		$12.3^{ab}\pm0.4$	$13.5^{a}\pm0.3$	

a, ... means on the same line with different superscripts are significantly different ($P < 0.05$)

al.(1976), Tilsch et al.(1977) and Harmsen and Westera (1981).

4 Conclusion

These preliminary results indicate that daily gain is hardly or not increased by crossbreeding with Charolais, Piedmont, White-Blue or Beefalo sires compared with purebred White-Red animals. This is due to the good growth capacity of the White-Red dual purpose breed. However, some of these crosses revealed a trend for a higher feed efficiency and a better carcass quality : higher dressing percent and/or lower fat content. There was nearly always a good relationship between the fat content in the carcass and the energy efficiency.

Acknowledgements

The authors wish to thank Ir. R. Moermans of the Biometric Unit of the CLO - Gent for the statistical analyses and Ir. R. Verbeke and Ir. G. Van de Voorde of the SKR at Melle for the dissection of the 8th rib cuts. They were also indebted to A. Hertegonne and R. Limpens for their skilled technical assistance.

References

Anonymous, 1975. Beefalo dress out at 62%, yield grades 1-2 in first major carcass data and cutting tests. Meat Industry, August, p. 22.

Bech Andersen, B., Liboriussen, T., Kousgaard, K. and Butcher, L. 1977. Crossbreeding experiments with beef and dual-purpose sire breeds on Danish dairy cows. III. Daily gain, feed conversion and carcass quality of intensively fed young bulls. Livest. Prod. Sci. 4, 19.

Boucqué, Ch.V., Cottyn, B.G., Aerts, J.V. and Buysse, F.X. 1976. Dried sugar beet pulp as a high energy feed for beef cattle. Anim. Feed Sci. Technol. 1, 643.

Boucqué, Ch.V., Fiems, L.O. and Buysse, F.X. 1977. Expérience acquise avec des vaches allaitantes de la race Charolaise au cours des trois premières lactations. Revue Agric., Brux. 30, 1105.

Boucqué, Ch.V., Fiems, L.O. and Buysse, F.X. 1978. Influence de l'alimentation de veaux d'élevage au lait reconstitué ou au pis, et du système alimentaire, après la période d'élevage, sur les performances de taureaux à viande. Revue Agric., Brux. 31, 255.

Boucqué, Ch.V., Fiems, L.O., Moermans, R.J., Cottyn, B.G. and Buysse, F.X. 1980. Effect of energy density of diets for intensive bull beef production on intake, growth rate and

feed conversion. Ann. Zootech. 29, 233, n° hors serie.
Brunnekreef, W. 1975. Uitkomsten geboorteregistratie en mestkalverproef van het Hendrix' proefkruisingsprogramma. Bedrijfsontwikkeling 6, 402.
Cunningham, E.P. 1974. Crossbreeding strategies in cattle populations. In : Proc. Work. Symp. on Breed Evaluation and Crossing Experiments with Farm Animals. 15-21 Sept. Zeist.
Fischer, W., Gribe, G. and Otto, E. 1976. Ergebnisse uber die Kreuzung verschiedener Fleischrinderrassen mit dem Schwarzbunten Rind. 2. Schlachtsleistung und Fleischbeschaffenheit von Romagnola und Piemonteser Kreuzungstieren. Arch. Tierzucht 19, 395.
Geay, Y. and Béranger, C. 1975. Aptitudes du Charolais à la production de viande. Bull. Tech. du C.R.Z.V. 19, 29.
Hanset, R. 1980. Muscular hypertrophy as a racial characteristic : the case of the Blue Belgian breed. EEC Seminar "Muscle Hypertrophy in Cattle", June 10-12th, Toulouse.
Harmsen, H.E. and Westera, A. 1981. Piemontese kruislingen brengen meer op. Bedrijfsontwikkeling 12, 57.
Kilkenny, J.B. 1975. Current practice in beef production from heifers. In : "The maiden female - a means of increasing meat production." Ed. J.B. Owen, School of Agric. Aberdeen.
Kilkenny, J.B. 1975. The economic implications of breed differences in commercial beef production systems. Winter meeting of the Irish Grassland and Animal Production Association.
Le Neindre, P. and Petit, M. 1976. Comportement maternel des bovins : application à l'adoption Ann. Méd. Vét. 120, 541.
Politiek, R.D. 1975. Mogelijheden van melk - en vleesproduktie bij kruisingen in de rundveeteelt. Bedrijfsontwikkeling. 6, 387.
Tilsch, K., Papstein, H.J., Otto, E., Mix, M. and Görlich, L. 1977. Mastund Schlachtleistungsergebnissen von Bullen aus der Gebrauchskreuzung des SR mit italienischen Fleischrindrassen. Tierzucht 31, 121.
Verbeke, R. and Van de Voorde, G. 1978. Détermination de la composition de demi-carcasses de bovins par la dissection d'une seule côte. Revue Agric. Brux. 31, 875.

CROSSBREEDING OF BEEF BULLS ON FRIESIAN COWS IN ITALY

A.L. Catalano

Istituto di Zootecnica Alimentazione e Nutrizione,
Facolta di Medicina Veterinaria, Universita di Parma,
Italy.

Summary

Charolais, Chianina, Marchigiana and Romagnola bulls were crossbred with Friesian cows.

Research was conducted over a two-year period and was concerned with conception rates, gestation period, calf birth weight, and difficulties in calving.

No change was found in conception rates, except for the Charolais-Friesian cross, in which conception was slightly reduced.

The Marchigiana-Friesian and Charolais-Friesian crosses had an increased gestation period.

The Chianina-Friesian and Romagnola-Friesian crosses had difficulties during calving, which were probably due to the high birth weight of the calves; the Romagnola-Friesian cross had the lowest mortality rate at birth.

Introduction

The first results of a crossbreeding experiment using beef bulls on Friesian cows, raised in the Province of Reggio Emilia, are presented in this paper.

This research was conducted with the collaboration of the "Consortium of Producers and Agricultural Co-operatives of Reggio Emilia", along with the assistance of local breeders and their associates.

The experiment was financed by the European Economic Community and the Italian Ministry of Agriculture and Forestry.

In 1976 and 1977 respectively 187 and 516 Friesian cows were artificially inseminated. The semen used came from four Charolais, four Chianina, five Marchigiana and seven Romagnola bulls. The cows used were those whose offspring were intended to be slaughtered and they were inseminated during the period between April and July.

The majority of the cows used were pluripara and in a few cases, primipara. All of the cows were fed a normal diet. The recorded data for each cow inseminated were :-

- the weight of each calf within 24 hours of birth;
- the difficulties in calving, including the necessity for veterinary assistance;
- the number of pre-natal, neo-natal deaths, and the number of abortions;
- the number of twin calvings;
- the duration of pregnancy was calculated from the day of insemination to the day of parturition.

Results

Table 1 shows the number of cows inseminated, the number and percentage of pregnancies, the number of births, the number of abortions and the gestation length of the cows.

The Table shows that in 1976 there was a relatively small number of inseminations from the Charolais bulls, thus the results from the two-year period are more valid.

The correlation between the age of the cow and the data will be presented in a more extensive paper.

Fertility: The difference among the Italian breeds in the number of pregnant cows is small, while there is an average difference of approximately 10% for the Charolais cross. There exists a uniformity in the percent of pregnant cows in each year, independent of the group, with the exception of the Charolais cross in 1976. The conception rate of the cow is probably more dependent on the season than on the breed of bull used.

Gestation period: The crossbreeding of purebred dairy cows with beef bulls caused an increase in the gestation length of the cow. This has also been documented by other Italian authors (Russo & Bazzani, 1970; Santoro *et al*.1975). In this research the crossbreeding of the Romagnola-Friesian had the shortest gestation length of 284.2 days. Other authors (Santoro *et al*. 1975) in similar research observed an average gestation length of 286.39 days.

Danish authors (Thysen, Liboriussen, & Bech Andersen, 1974) crossing Romagnola bulls with Black-spotted, Red-spotted Danish

Table 1 Cows

Breed of bull	Year	Inseminated No.	Pregnant No.	Pregnant %	Births No.	Abortions No.	Length of gestation (days)
Charolais	1976	29	16	55.17	16	-	286.62 ± 3.10
	1977	145	84	57.93	81	3	286.81 ± 3.86
	1976/77	174	100	57.49	97	3	286.78 ± 3.74
Chianina	1976	89	71	79.78	69	2	285.14 ± 4.37
	1977	179	104	58.10	101	3	284.63 ± 4.87
	1976/77	268	175	66.90	170	5	284.84 ± 4.68
Marchigiana	1976	34	26	76.47	24	2	288.21 ± 3.40
	1977	89	52	58.43	51	1	287.76 ± 4.46
	1976/77	123	78	64.44	75	3	287.91 ± 4.51
Romagnola	1976	35	27	77.14	27	-	284.41 ± 4.80
	1977	103	60	58.25	59	1	284.10 ± 4.51
	1976/77	138	87	64.11	86	1	284.20 ± 4.60

and Jersey cows, observed an average gestation length of 286 days.

The cross Chianina-Friesian had a slightly longer gestation length of 284.8 days, and for the same cross the Danish authors in their experiment observed a gestation length of 288.1 days. In the cross Marchigiana-Friesian, the gestation length was longest, 287.9 days.

Comparing this data with the normal gestation length for pure Friesians of 279.4 days (Santoro, et al., 1975) this research showed an increased gestation length of 4.76 days for the Romagnola-Friesian cross, 5.44 days for the Chianina-Friesian cross, 7.32 days for the Charolais-Friesian cross and 8.49 days for the Marchigiana-Friesian cross. It is interesting to note that there was no significant difference in gestation length in the two years of the experiment.

Calving difficulties: Table 2 shows the classification of calvings according to the following criteria :-

- easy delivery, required only the presence of the farmer;
- some difficulty, classified by the necessity of having more than one person assisting;
- very difficult, requiring at least three people;
- most difficult, requiring veterinary assistance.

The data showed that the Romagnola-Friesian cross had the most frequent number of calving problems with an incidence of 15.12% in 86 births, but they also had the lowest number of deaths among the calves - only one.

For the other crosses, the incidence of calving problems was :-

- 9.33% for the Marchigiana;
- 8.25% for the Charolais;
- 10.59% for the Chianina.

The Chianina cross had the greatest need for veterinary assistance (10.6%). This is explained by the particular structure of the calf, caused by this crossbreeding, and its high birth weight.

The other crossbreds each needed veterinary help 9.30% of the time. There was less difficulty observed in the Marchigiana Friesian cross, with 49.34% having easy deliveries and only 32% requiring the help of two people.

Table 2 Calving difficulties

Breed of bull	Year	Births No.	Easy delivery No.	Easy delivery %	Some difficulty No.	Some difficulty %	Very difficult No.	Very difficult %	Most difficult No.	Most difficult %
Charolais	1977	16	9	56.24	3	18.76	2	12.5	2	12.5
	1978	81	32	39.50	36	44.45	6	7.41	7	8.64
	1977/78	97	41	42.27	39	40.21	8	8.25	9	9.27
Chianina	1977	69	32	46.38	23	33.33	6	8.69	8	11.60
	1978	101	26	25.74	53	52.48	12	11.88	10	9.90
	1977/78	170	58	34.12	76	44.70	18	10.50	18	10.60
Marchigiana	1977	24	15	62.50	5	20.83	1	4.17	3	12.50
	1978	51	22	43.14	19	37.25	6	11.76	4	7.85
	1977/78	75	27	49.34	24	32.00	7	9.33	7	9.33
Romagnola	1977	27	14	51.85	8	29.63	3	11.11	2	7.41
	1978	59	20	33.90	23	38.98	10	16.95	6	10.17
	1977/78	86	34	39.53	31	36.05	13	15.12	8	9.30

The Marchigiana-Friesian cross seemed to have few problems in calving, but when there were problems, they were usually serious. This was also true for the Charolais.

In the second year of research it was noticed that the presence of a veterinarian was required less often among all breeds, except the Romagnola cross. It is thought that this was due to the experience which the farmers gained in the previous year. This experience is important in that it reduces the need for outside help, and lowers production costs.

Reggio Emilia (the research zone) is a Province with mostly dairy cows; the farmers lack experience in working with the birth of beef cattle, and thus the real difficulties in calving were probably exaggerated the first year and in many cases more people were present than were actually needed.

It should also be noted that some of the cows in the experiment were primipara and this caused an increased number of calving difficulties. This will be further investigated in future research.

Birth weight: Birth weight is a factor of heredity, as is demonstrated from many observations. The weight at birth is presented in Table 4, which shows the following average results :-

- 42.23 kg among 102 Charolais-Friesian cross births;
- 42.62 kg among 78 Marchigiana-Friesian cross births;
- 45.10 kg among 87 Romagnola-Friesian cross births;
- 47.35 kg among 170 Chianina-Friesian cross births.

Similar research conducted on the crossing of Romagnola bulls and Friesian cows showed an average weight of 43.56 kg among 30 births (Santoro *et al.*, 1975).

The males had an average birth weight greater than the females, which varied from 3.7 kg in Chianina-Friesian crosses to 2.8 kg with the Marchigiana-Friesian crosses, and 2.1 kg with the Romagnola-Friesian crosses.

The small difference in weight between the males and females born of the Charolais bull, we feel is due to the birth of three pairs of male twins and one set of male triplets; in fact, if we calculate the average weight of the male at birth, excluding weights of the twins and triplets, we arrive at the median weight of 45.4 kg which corresponds to the weight found

Table 3 Calves born

Breed of bull	Year	Total	Alive		Dead		Dead after birth		Multiple births	
			No.	%	No.	%	No.	%	No.	%
Charolais	1977	17	16	94.12	1	5.88	2	12.5	1	6.25
	1978	85	84	98.82	1	1.18	4	4.76	2+1 (tripl)	3.70
	1977/78	102	100	98.04	2	1.96	6	6.00	4	4.12
Chianina	1977	69	67	97.10	2	2.90	1	1.49	-	-
	1978	101	97	96.04	4	3.96	3	3.09	-	-
	1977/78	170	164	96.47	6	3.53	4	2.44	-	-
Marchigiana	1977	25	24	96.00	1	4.00	3	12.5	1	4.17
	1978	53	51	96.08	2	3.92	2	3.92	2	3.92
	1977/78	78	75	96.05	3	3.95	5	6.67	3	4.00
Romagnola	1977	27	27	100	-	-	-	-	-	-
	1978	60	60	100	-	-	1	1.67	1	1.69
	1977/78	87	87	100	-	-	1	1.15	1	1.16

Table 4 Birth weight of calves, by sex

Breed of bull	Year	Males			Females			Total		
		No.	Kg	(1)	No.	Kg	(1)	No.	Kg	(1)
Charolais	1977	13	41.4 ± 6.0	(43.8)	4	44.0 ± 1.9	-	17	42.0	(43.8)
	1978	39	42.5 ± 7.8	(45.9)	46	42.0 ± 3.6	-	85	42.2	(43.8)
	1977/78	52	42.2 ± 7.4	(45.4)	50	42.1 ± 3.5	-	102	42.2	(43.8)
Chianina	1977	40	49.3 ± 3.0	-	29	45.1 ± 3.5	-	69	47.91	-
	1978	53	48.9 ± 3.3	-	48	45.6 ± 3.5	-	101	47.34	-
	1977/78	93	49.1 ± 3.1	-	77	45.4 ± 3.5	-	170	47.35	-
Marchigiana	1977	12	44.6 ± 3.0	-	13	40.8 ± 5.3	(42.9)	25	42.62	(43.7)
	1978	30	43.7 ± 4.4	(44.6)	23	41.2 ± 5.0	(42.4)	53	42.61	(43.6)
	1977/78	42	43.9 ± 4.0	(44.6)	36	41.1 ± 5.1	(42.6)	78	42.62	(43.7)
Romagnola	1977	14	45.9 ± 2.8	-	13	44.0 ± 2.7	-	27	44.99	-
	1978	25	46.5 ± 4.8	(47.0)	35	44.2 ± 3.8	(44.6)	60	45.16	(45.6)
	1977/78	39	46.3 ± 4.2	(46.6)	48	44.2 ± 4.0	(44.4)	87	45.10	(45.4)

(1) Average weight of calves excluding weights of multiple births

by other researchers.

Conclusions

From the data obtained in our research, we reached the following conclusions :-

1) In the crossbreeding of the various beef breeds with Friesian, there was no significant difference with regard to conception rates between the various breeds, and particularly the Italian breeds. There is no decrease in conception rates with crossbreeding.

2) Crossbreeding with beef breeds results in an increased gestation period. The increase in gestation time is different for each breed. The Romagnola-Friesian cross had the shortest gestation period and the Marchigiana had the longest.

3) The Charolais-Friesian and Marchigiana-Friesian crosses have the lowest average birth weight and thus have less difficulty with normal deliveries, but sustain higher mortality rates when there are problems. The Romagnola-Friesian and Chianina-Friesian crosses show a higher birth weight, and thus have more difficulty at calving time, especially among primipara cows, but at the same time the Romagnola have the lowest incidence of deaths at birth, while in contrast, the Chianina have the higher death rate and not one multiple birth.

References

Belic, M. and Menissier, F. 1968. Étude de quelques facteurs influençant les difficultés de velage en croisement industriel. Ann. Zootech. 17, 107.

Borghese, A., Romita, A. and Gigli, S. 1977. Caratteristiche produttive di meticci Frisoni in confronto con la razza pura. Accrescimento e resa al macello. Ann. 1st. Sper. Zootec. 10, 2, 161-162.

Crowley, J.P. 1965. The effect of Charolais bulls on calving performance. Irish J. agric. Res. 4, 285.

Naude, R.T. 1967. Birth weight of pure and crossed dairy calves with observations on the ease of parturition of their dams. Proc. S. Afr. Anim. Prod. 137.

Russo, V. and Bazzani, D. 1970. Osservazioni sul peso alla nascita del vitello sulla durata della gestazione e sul comportamento al parto nell'incrocio Charolais-Frisona. Relazione al 3° Symposium Nazionale sulla Genetica Animale. Asti, 11-12 maggio, 1970.

Santoro, P., Monetti, P.G., Zaghini, G and Parisini, P. 1975. L'incrocio industriale Romagnola-Frisona nella produzione della carne bovina. Rivista di Zootecnia e Veterinaria, 37, 37-39.

Thysen, I., Liboriussen, T. and Bech Andersen, B. 1974.Kridsnings og produktionstineforsog med europaeiske kodracer. Meddelelse, 4.

Turton, J.D. 1964. The Charolais and its use in crossbreeding. Anim. Breed. Abstr. 32, 119.

DISCUSSION ON DR. L.O. FIEMS'S AND DR. L. CATALANO'S PAPERS

Chairman: Dr. R. Thiessen

Langholz: It puzzles me that Marchigiana crosses showed the longest gestation length and the lowest birth weight. Can it be that there was a confounding of sire breed with farm effect? In parallel studies with Marchigiana crosses in Germany, not only did they have the longest gestation lengths, but also the highest birth weights.

Catalano: Perhaps, but I do not think so. There were about 20 breeders involved, but only 5 bulls were sampled for the Marchigiana breed.

Cunningham: In both experiments both male and female offspring were included. Do the breed types rank similarly when evaluated as males and females, in other words, was there a breed by sex interaction ?

Fiems: The results are very preliminary and I cannot draw firm conclusions.

Menissier: While it is possible that an interaction of breed with sex occurred, it is more likely that the interaction was between breed and feeding system.

FATTENING PERFORMANCE AND SLAUGHTER TRAITS OF PUREBRED AND CROSSED YOUNG BULLS*

M. Bonsembiante, G. Bittante, P. Cesselli, I. Andrighetto

Istituto di Zootecnica, Universita degli Studi di Padova, Via Gradenigo, 6 - 35100 Padova, Italy.

Summary

In order to evaluate the effect of crossing beef bulls with dual-purpose cows on the performance of fattening young bulls, a trial was carried out on 74 calves of the following genetic types : Italian Brown Swiss (IBS), German Simmental (GS) and crosses obtained from Piedmont (PD x IBS and PD x GS)Limousin (LM x IBS and LM x GS) and Charolais (CH x GS) bulls.

Results showed that whilst the actual dam breed has no great effect on the performances of the young bulls, crossing with beef breeds did improve the performance of the IBS and GS as regards dressing percentage, the incidence of the fifth quarter and the rib eye area.

The use of Piedmont bulls resulted in a reduction of the daily gain with an accompanying increase of dry matter conversion and a marked improvement of dressing percentage, as the weights of the hide, lights and perivisceral fatty deposits were reduced. The crossing with Piedmont also increased the percentage of lean meat and bone, and reduced the incidence of fatty deposits in the sample joint taken from the tenth rib.

The Limousin crosses were characterised by a reduced intake of dry matter, but with a good conversion, a high dressing percentage and a smaller proportion of bone in the sample joint.

The CH x GS crosses showed a good growth rate, possessing slaughter traits which were midway between those of the PD and LM crosses, and those of the two dual-purpose IBS and GS breeds.

Introduction

The majority of cattle bred in Europe belong to dual-purpose breeds. The integrated production of meat and milk is widespread, and has not been replaced by productive systems which specialise only in meat or milk, such as can be found in other Continents.

* This work was supported by the National Research Council of Italy (Progetto Finalizzato: 'Incremento disponibilità alimentari di origine animale').

In Europe, we have seen an improvement in milk productivity in respect to that of meat. This trend is, however, in contrast to the needs of the consumer, who is continually asking for more beef. One solution to this problem would be to reduce the number of dairy cattle and increase the breeding of beef cattle, but this could only be a partial solution due to the structural and economic limits involved. The confined breeding of suckling cows based on the utilisation of agricultural and industrial by-products would also alleviate the problem (Bonsembiante & Chiericato, 1977; Bonsembiante et al., 1981). Yet another important contribution to the supply of beef could be made by improving and increasing the amount of beef obtainable from dairy herds, particularly from dual-purpose cattle. The crossing of beef bulls is one of the most valid means of achieving this end.

The object of the trial described here was to evaluate the improvement in performance of young bulls obtained from the crossing of Piedmont, Limousin and Charolais bulls with Brown Swiss and Simmental cows, these being the most widespread dual-purpose breeds in Europe.

Materials and Methods

The trial was carried out on the Experimental Farm of the Faculty of Agricultural Science of the University of Padua at Legnaro, on 74 calves of the following genetic types :-

9 Italian Brown Swiss (IBS)
12 Piedmont x IBS
12 Limousin x IBS
9 German Simmental (GS)
9 Piedmont x GS
10 Limousin x GS
13 Charolais x GS

The experiment began when the calves had reached an average weight of 176 kg.

The feeding programme consisted of an *ad lib.* diet of maize grain and maize stover silage at a ratio of 3 : 1 on a dry basis, and 1 kg per day of protein supplement.

The chemical composition of the feedstuffs was determined by the methods of the A.O.A.C. (1970) and the percentage composition of the supplement is reported in Table 1. The metabolisable energy content was calculated using the equations of Nehring and Haenlein (1973); the digestibility coefficients for maize grain and maize stover silage were reported by Bonsembiante (1981) and those for the soybean meal of the supplement by Morrison (1957).

The energy balance of the animals on trial was estimated using the values of Net Energy for maintenance (NEm) and of Net Energy for growth (NEg) of the diet estimated according to the method of Lofgreen and Garrett (1968).

The consumption of feedstuffs was controlled daily for each of the 28 groups of animals bred (4 replications for each of the 7 genetic types).

The liveweight of the young bulls was controlled monthly and various live-animal measurements were carried out at the beginning and at the end of the trial. The animals were slaughtered at an average liveweight of 548 kg according to the technique of the Institute of Zootecnica of Padua (Bonsembiante, 1962). After slaughter the carcasses were measured according to the method suggested by De Boer et al. (1974) and the 10th rib sample joint was removed for determination of the incidence of lean meat, fat and bone, and for the measurement of the rib eye area (Longissimus dorsi section) (Lanari, 1973).

The statistical analysis of the data not only takes into account the effect of the genetic type but also two other completely crossed factors: that of the physical form of the maize grain (50% of the animals received whole shelled grain; 50% maize meal); the other factor was the fattening barn (barns A and B).

For group data (feed intake, feed conversion and energy balance) the statistical significance was tested on the triple interaction (genetic type x physical form of the maize x type of barn). The principal effect of the genetic type is broken down in the orthogonal comparisons shown in Tables 2 to 8 (Snedecor & Cochran, 1967).

Table 1 Chemical composition and nutritive value of feedstuffs employed in the trial (mean ± s.d.)

	Maize grain	Maize stover silage	Supplement*
Sample analysed	40	22	18
Moisture %	13.7 ± 1.8	64.9 ± 5.8	14.5 ± 2.5
Crude protein % DM	9.4 ± 0.5	5.1 ± 0.6	40.0 ± 1.9
Ether extract % DM	4.0 ± 0.4	0.8 ± 0.2	1.2 ± 0.4
Crude fibre % DM	2.7 ± 0.6	34.6 ± 2.3	4.4 ± 0.9
Ash % DM	1.4 ± 0.1	14.8 ± 2.8	26.4 ± 2.9
N-free extracts % DM	82.5 ± 0.9	44.7 ± 2.1	28.0 ± 2.1
pH		6.3 ± 0.5	
Metabolisable energy MJ/kg DM	12.9	6.3	9.9

* Composition of the supplement: soybean meal solv. ext. 66.5%; limestone 15.0%; sodium acid phosphate 15.0%; salt 3.0% and vitamin-trace mineral premix 0.5%.
Integrated with: vitamin A 42,000 IU/kg; vitamin D_3 2,800 IU/kg; vitamin E 28 ppm; vitamin K 7 ppm; Zn 259 ppm; Mn 132 ppm; Cu 38 ppm; Fe 24 ppm; Co 2.4 ppm and I 1.4 ppm.

Results

As regards the purebred Brown Swiss and Simmental, the daily gain of the crossed young bulls was significantly influenced by the sire breed (Table 2). Whereas the use of Charolais bulls improved the growth rate, the opposite effect occurred when Piedmont bulls were employed. No difference was, however, noted with Limousin crosses with respect to purebred young bulls.

The intake of dry matter was also affected by the genetic type of the animal (Table 3). The use of Limousin bulls caused a significant reduction in the consumption of dry matter by their offspring, whilst when Piedmont bulls were used, this effect was only noted when the dams were Simmental. The Charolais x Simmental crosses, on the contrary, possessed good appetites, like their purebred counterparts.

The dry matter conversion and the ratio between the Net Energy available for the growth and daily gain achieved were thus significantly lower for the crosses with French beef breeds, whereas higher values were achieved by offspring of Piedmont bulls.

The conformation of the animal was significantly influenced by genetic type, as can be seen from the live animal measurements, taken at the beginning and the end of the trial, co-varied with the liveweight (Table 4).

At the beginning of the trial, the Piedmont and the Limousin crosses were higher at the withers than the Brown Swiss and the Simmental, and at the end of the trial the Piedmont crosses were still higher than the other bulls. The trunk length of the Charolais x Simmental crosses was somewhat greater at the beginning, whilst before slaughter it was noted that only the Piedmont crosses were longer in this respect than the Limousin. On the other hand, Limousins possessed a greater chest girth at the end of the trial. Thus the Limousins proved to be shorter and more compact than the offspring of Piedmont bulls.

A significant effect of crossbreeding on dressing percentage was noted at slaughter (Table 5). Both the dressing percentages on live slaughter weight and on empty bodyweight improved when Piedmont and Limousin bulls were crossed with Brown Swiss and Simmental dams, whilst the effect of Charolais bulls on these

Table 2 Liveweight and daily gain

	No.	Initial liveweight (kg)	Final liveweight (kg)	Length of trial (days)	Daily gain g/d
1. Italian Brown Swiss (IBS)	9	179.8	542.4	365.5	1,001
2. Piedmont x IBS	12	181.0	536.2	379.5	942
3. Limousin x IBS	12	182.9	539.3	363.5	982
4. German Simmental (GS)	9	172.5	556.2	380.8	1,016
5. Piedmont x GS	9	189.4	540.1	376.3	936
6. Limousin x GS	10	169.0	552.3	379.0	1,021
7. Charolais x GS	13	167.9	542.5	353.8	1,061
Overall	74	177.5	544.2	371.0	994
Error mean square		478.28	1,416.48	1,951.63	8,608
Orthogonal comparisons:					
- dam breed effect (1 + 2 + 3 vs 4 + 5 + 6)		NS	NS	NS	NS
- crossbreeding effect (2 + 3 + 5 + 6 vs 1 + 4)		NS	NS	NS	NS
- Piedmont vs Limousin (2 + 5 vs 3 + 6)		NS	NS	NS	$P < .05$
- Charolais effect (7 vs 1 + 2 + 3 + 4 + 5 + 6)		NS	NS	NS	$P < .05$
- other comparisons		NS	NS	NS	NS

NS = not significant

Table 3 Daily intake and conversion of dry matter and ratio between net energy available for growth (N.E. g) and weight gain

	Dry matter daily intake			Dry matter conversion	N.E. g
	kg per head	% Liveweight	g/kg $\text{Liveweight}^{0.75}$		Wt. gain Mj/kg
1. Italian Brown Swiss (IBS)	7.29	2.02	87.9	7.32	15.6
2. Piedmont x IBS	7.31	2.04	88.6	7.79	16.7
3. Limousin x IBS	6.77	1.88	81.7	6.90	13.4
4. German Simmental (GS)	7.33	2.01	87.9	7.27	15.5
5. Piedmont x GS	7.07	1.94	84.7	7.60	15.5
6. Limousin x GS	7.04	1.95	85.0	6.96	14.2
7. Charolais x GS	7.22	2.03	88.2	6.81	14.6
Overall	7.14	1.98	86.3	7.24	15.1
Error square mean	0.1415	0.0012	4.54	0.5142	2.12
Orthogonal comparisons:					
- dam breed effect (1+2+3vs4+5+6)	NS	NS	NS	NS	NS
- crossbreeding effect (2+3+5+6vs1+4)	NS	$P < .01$	$P < .05$	NS	NS
- Piedmont vs Limousin (2+5vs3+6)	NS	$P < .01$	$P < .01$	$P < .05$	$P < .05$
- Charolais effect (7vs1+2+3+4+5+6)	NS	$P < .05$	NS	NS	NS
- Other comparisons	NS	$P < .01$	$P < .05$	NS	NS

NS = not significant

Table 4 Live animal measurements at the beginning and at the end of the trial (cm)

	Height at withers		Length of trunk		Chest girth	
	Initial	Final	Initial	Final	Initial	Final
1. Italian Brown Swiss (IBS)	98.5	125.4	108.0	141.8	133.7	193.1
2. Piedmont x IBS	101.4	127.1	107.8	147.0	134.1	193.6
3. Limousin x IBS	100.2	123.4	106.3	143.0	133.2	196.9
4. German Simmental (GS)	98.3	125.3	107.5	141.6	134.3	197.8
5. Piedmont x GS	100.3	127.5	105.7	142.8	134.9	192.7
6. Limousin x GS	100.0	125.8	109.1	139.5	136.4	197.2
7. Charolais x GS	99.7	125.5	110.8	143.1	132.6	194.0
Overall	99.8	125.7	107.9	142.7	134.2	194.9
Error mean square	4.40	9.03	8.98	24.79	9.87	16.62
Orthogonal comparisons:						
- dam breed effect (1+2+3vs4+5+6)	NS	NS	NS	NS	NS	NS
- crossbreeding effect (2+3+5vs1+4)	P <.01	NS	NS	NS	NS	NS
- Piedmont vs Limousin (2+5vs3+6)	NS	P <.01	NS	P <.05	NS	P <.05
- Charolais effect (7vs1+2+3+4+5+6)	NS	NS	P <.01	NS	NS	NS
- other comparisons:	NS	NS	NS	NS	NS	P <.05

NS = not significant

Table 5 Slaughter, empty body and cold carcass weights and dressing percentages

*	Slaughter live wt. kg	Empty body wt. kg	Cold carcass wt. kg	Dressing percentage	
				On S.L. wt. %	On E.B. wt. %
1. Italian Brown Swiss (IBS)	544.5	485.3	318.0	58.4	65.5
2. Piedmont x IBS	540.3	491.0	332.2	61.5	67.7
3. Limousin x IBS	546.8	492.5	331.8	60.8	67.3
4. German Simmental (GS)	560.8	506.7	329.7	58.8	65.0
5. Piedmont x GS	539.2	486.2	327.1	60.7	67.3
6. Limousin x GS	555.6	499.2	337.9	60.8	67.7
7. Charolais x GS	547.8	490.1	325.0	59.3	66.2
Overall	547.8	493.1	328.9	60.0	66.7
Error mean square	1,491.59	1,227.57	699.91	2.87	2.36
Orthogonal comparisons:					
- dam breed effect (1+2+3vs4+5+6)	NS	NS	NS	NS	NS
- crossbreeding effect (2+3+5+6vs1+4)	NS	NS	NS	$P < .01$	$P < .01$
- Piedmont vs Limousin (2+5vs3+6)	NS	NS	NS	NS	NS
- Charolais effect (7vs1+2+3+4+5+6)	NS	NS	NS	NS	NS
- Other comparisons	NS	NS	NS	NS	NS

N S = not significant

variables was much less evident.

The percentage incidence of the hide on empty bodyweight was much more favourable for crosses, particularly Piedmont offspring, whilst crossbreeding had much less effect on the other components of the 5th quarter (Table 6).

The only significant effect noted in the entire trial between the two dam breeds concerned the size of the distal fore and hind legs, which was greater in the Brown Swiss purebred and crosses than in the Simmental. The Charolais x Simmental crosses possessed a digestive tract which was considerably heavier, this being related to their good dry-matter intake. Lastly, the Piedmont crosses, particularly those crossed with Brown Swiss, showed a modest amount of perivisceral fat.

Unlike the live animal conformation, the carcass conformation was in no way affected by the genetic type (Table 7, showing carcass measurements co-varied for weight).

Judging from the results of the dissection of the sample joint (Table 8) the tissue composition of the carcass was, on the contrary, strongly influenced by the genetic type of the young bulls. The offspring of the Piedmont bulls, especially those with Brown Siwss dams, demonstrated a greater incidence of lean meat and bone and a lower incidence of fatty deposits, not only in comparison with Limousin crosses, but also with other subjects.

Lastly, crossing with beef breeds resulted in a clear and significant muscular hypertrophy, as was shown by the increase of the rib eye area; this was particularly evident when the dam breed was Brown Swiss.

Discussion and Conclusion

The dual-purpose Brown Swiss and Simmental calves demonstrated satisfactory productive performance until they reached a liveweight which, in relation to their feeding programme, is to be considered rather high. Fatness was not excessive, and was in fact similar to that of animals which had been slaughtered at a lower weight (Bonsembiante *et al.*, 1974; Rioni, Bittante & Susmel, 1979; Bonsembiante & Bittante, 1981).

Table 6 Percentage incidence of the components of the fifth quarter on empty body weight

	Hide	Head	Feet	Lights*	Empty digestive tract	Peri-visceral fat
1. Italian Brown Swiss (IBS)	11.73	3.89	2.10	3.87	5.42	1.51
2. Piedmont x IBS	10.66	3.93	2.05	3.54	5.09	1.13
3. Limousin x IBS	11.34	3.80	2.01	3.68	5.31	1.63
4. German Simmental (GS)	12.19	3.74	2.03	3.70	5.33	1.38
5. Piedmont x GS	10.82	3.96	1.98	3.67	5.16	1.64
6. Limousin x GS	11.39	3.71	1.88	3.64	5.23	1.64
7. Charolais x GS	11.78	3.89	2.03	3.62	5.71	1.49
Overall	11.42	3.85	2.01	3.68	5.32	1.46
Error mean square	0.6145	0.0444	0.0233	0.0563	0.1267	0.1117
Orthogonal comparisons:						
- dam breed effect (1+2+3vs4+5+6)	NS	NS	$P<.05$	NS	NS	NS
- crossbreeding effect (2+3+5+6vs1+4)	$P<.01$	NS	NS	$P<.05$	NS	NS
- Piedmontvs Limousin (2+5vs3+6)	$P<.05$	$P<.01$	NS	NS	NS	$P<.01$
- Charolais effect (7vs1+2+3+4+5+6)	NS	NS	NS	NS	$P<.01$	NS
- other comparisons	NS	NS	NS	Ns	NS	NS

NS = not significant

* Lungs, trachea, heart, spleen, diaphragm and liver

Table 7 Carcass measurements (cm)

	Length of carcass	Depth of chest	Length of leg	Width of leg
1. Italian Brown Swiss (IBS)	128.9	41.6	65.8	28.0
2. Piedmont x IBS	131.6	41.0	67.1	29.8
3. Limousin x IBS	128.2	39.8	67.4	29.6
4. German Simmental (GS)	128.3	41.0	66.5	30.3
5. Piedmont x GS	129.0	41.2	67.0	29.0
6. Limousin x GS	130.3	41.2	66.1	29.6
7. Charolais x GS	130.8	40.9	67.5	29.4
Overall	129.6	41.0	66.8	29.4
Error mean square	14.67	2.10	6.50	1.59
Orthogonal comparisons:				
- dam breed effect (1+2+3vs4+5+6)	NS	NS	NS	NS
- crossbreeding effect (2+3+5+6vs1+4)	NS	NS	NS	NS
- Piedmont vs Limousin (2+5vs3+6)	NS	NS	NS	NS
- Charolais effect (7vs1+2+3+4+5+6)	NS	NS	NS	NS
- other comparisons	NS	NS	NS	$P < .05$

NS = not significant

Table 8 Tissue composition of the 10th rib sample joint and rib eye area

	Lean %	Fat %	Bone %	Rib eye area cm^2
1. Italian Brown Swiss (IBS)	69.4	14.1	16.5	78.8
2. Piedmont x IBS	74.1	8.6	17.3	87.5
3. Limousin x IBS	69.7	14.3	16.0	89.5
4. German Simmental (GS)	68.3	13.4	18.3	82.6
5. Piedmont x GS	70.0	11.4	18.7	87.0
6. Limousin x GS	69.5	15.3	15.2	85.2
7. Charolais x GS	69.6	14.6	15.8	85.7
Overall	70.1	13.1	16.9	85.2
Error mean square	12.10	9.68	6.74	87.45
Orthogonal comparions:				
- dam breed effect (1+2+3vs4+5+6)	NS	NS	NS	NS
- crossbreeding effect (2+3+5+6vs1+4)	NS	NS	NS	$P < .05$
- Piedmont vs Limousin (2+5vs3+6)	$P < .05$	$P < .01$	$P < .01$	NS
- Charolais effect (7vs1+2+3+4+5+6)	NS	NS	NS	NS
- other comparisons	NS	NS	NS	NS

NS = not significant

The crossing of cows from the above-mentioned breeds with beef bulls does not improve the rat e of growth or nutritonal efficiency as much as it does the slaughter and dissection traits, thus confirming the results reported by Leuenberger et al., 1977; by Tilsch, Otto & Papstein, 1977; and by Fercej & Osterc, 1978.

Both purebreds and crosses proved to be very efficient in terms of nutritional efficiency, having required an amount of NEg per kilo of daily gain 20-36% less than the requirements reported by Lofgreen and Garrett (1968) for steers of equal weight belonging to Anglo-Saxon breeds.

Such a result can be attributed to the different ratio between production of muscle and fatty deposits which characterise these breeds.

Regarding the sire breed, once again the Piedmont has shown that it is the breed which maximises the ratio between lean meat and fat and which improves the dressing percentage, at the same time confirming the modest growth rate which betrays a not especially efficient feed conversion (Bonsembiante, Rioni & Chiericato, 1975; Casu et al., 1975; Bonsembiante, Chiericato & Lanari, 1976).

The Limousin demonstrated a greater degree of fatness and a dressing percentage which was as high as that of the Piedmont crosses, and also an improved nutritional efficiency, thus confirming the many results obtained from home and abroad.

It is generally accepted that the Charolais, when used as a sire breed, has a positive effect on growth rate, even when the dam breed is characterised by good weight gain. Slaughter performance was, however, lower than that of Piedment and Limousin, even though the Charolais does have a positive influence on the colour of the meat (Bittante & Andrighetto, 1981).

In conclusion, the crossing of beef bulls with dual-purpose dams, as with dairy cows, has proved to be a technique of reproduction which results in an improved production of beef. It would seem clear that the choice of sire breed depends on the importance attributed to such factors as growth rate, nutritional efficiency, dressing percentage, carcass composition and the quality of the meat. Other factors, however, must not be forgotten, such as : difficulties at calving, resistance to disease, the specific demands of the market and the type of production.

References

A.O.A.C. 1970. Official Methods of Analysis. A.O.A.C. Washington, DC.

Bittante, G. and Andrighetto, I. 1981. The effect of freezing on the meat colour of purebred and crossed young bulls. EEC Seminar on Beef Production from the Dairy Herd, Dublin.

Bonsembiante, M. 1962. Contributo sperimentale allo studio della produzione del vitellone di razza rossa friulana. Riv. Zoot. 35, 302.

Bonsembiante, M. 1981. Unpublished data.

Bonsembiante, M. and Bittante, G. 1981. Produzione del vitellone: confronto fra tipi genetici diversi (Limousin, Charolais, Charolais x Aubrac, Saler, Simmental Bavarese, Bruno Alpino, Frisone Polacco e Frisone Inglese. In press.

Bonsembiante, M. and Chiericato, G.M. 1977. Allevamento della manza in ambiente confinato per la produzione del vitello: effetto di differenti fonti di integrazione azotata della razione. Zoot. e Nutr. Anim. 3, 41.

Bonsembiante, M., Chiericato G.M. and Lanari, D. 1976. Current research on commercial crossbreeding for beef in dairy herds in Italy. EEC Seminar on crossbreeding experiments and strategy of beefutilisation to increase beef production. Verden.

Bonsembiante, M., Lanari, D., Susmel, P. and Andrighetto, I. 1981. Produzione del vitello in ambiente confinato con manze B.A. e P.R.B. 2) Effetto della somministrazione di silomais e di stocchi di mais insilati sulla gravidanza e sul parto. Zoot. e Nutr. Anim. In press.

Bonsembiante, M., Rioni, M. and Chiericato, G. M. 1975. Risultati sperimentali dell'incrocio fra bovini di razze italiane da carne e-da latte. Genetica Agraria, XXIX, n. 1/2, 33.

Bonsembiante, M., Rioni, M., Chiericato, G.M. and Susmel, P. 1974. Incrocio fra tori di razze da carne e vacche da latte per la produzione del vitellone. Riv. Zoot. Vet. 5, 383.

Casu, S., Boyazoglu, J.G., Bibe, B. and Vissac, B. 1975. Systèmes d'amelioration génétique de la production de viande bovine dans les pays mediterraneens : les recherches sardes. Bull. tech. Dep. Genet. Anim. (INRA), 22.

De Boer, H., Dumont, B.L., Pomeroy, R.W. and Weniger, J.H. 1974. Manual on EAAP reference methods for the assessment of carcass characteristics in cattle. Livest. Prod. Sci. 1, 151.

Fercej, J. and Osterc, J. 1978. L'incrocio interrazziale di bestiame bruno alpino con razze specializzate da carne. 1st World Conf. of Brown Swiss Breeders, Innsbruck.

Lanari, D. 1973. Utilizzazione dei tagli campione nella stima

della composizione delle carcasse bovine. Riv. Zoot. e Vet. n. 3, 241.
Leuenberger, H., Schneeberger, M., Gaillard, C., Hanser, H., Kunzi, N. and Weber, F. 1977. Gebrauchskreuzungen mit Fleischrassen Ergebnisse und vorläufige Schlussfolgerungen. Schweizerische Landwirtschaftliche Monatshefte, 55, 138.
Lofgreen, G.P. and Garrett, W.N. 1968. A system for expressing net energy requirements and feed values for growing and finishing beef cattle. J. Anim. Sci. 27, 793.
Morrison, F.B. 1957. Feeds and Feeding. The Morrison Publ. Co. Ithaca, New York.
Nehring, K. and Haenlein, G.F.W. 1973. Feed evaluation and ration calculation based on net energy fat. J. Anim. Sci. 36, 949.
Rioni, M., Bittante, G. and Susmel, P. 1979. Produzione del vitellone: confronto tra tipi genetici diversi (Limousine, Charolaise, Guascona, Pezzata Rossa, Normanna, Bruna Alpina e Frisona). Zoot. e Nutr. Anim. 5, 565.
Snedecor, G.W. and Cochran, W.G. 1967. Statistical Methods. Iowa State University Press, Ames, Iowa.
Tilsch, K., Otto, E. and Papstein, H.J. 1977. Ergebnisse der Genotypenprüfung von Masthybriden aus italienischen Fleischrinderrassen. Arch. Tierzucht. 20. 257.

THE EFFECT OF FREEZING ON THE MEAT COLOUR OF PUREBRED AND CROSSED YOUNG BULLS

G. Bittante I. Andrighetto

Istituto di Zootecnica, Universita degli Studi di Padova,
Via Gradenigo, 6 - 35100 Padova, Italy.

Summary

The moisture, fat and pigment contents, and the lightness (L), redness (a_L), yellowness (b_L), saturation (S) and hue (H) of samples of Longissimus dorsi muscle from 67 young bulls of the following genetic types were determined : Italian Brown Swiss (IBS), German Simmental (GS), Piedmont (PD) x IBS, PD x GS, Limousin (LM) x IBS, LM x GS and Charolais (CH) x GS.

The physical variables L, a_L, b_L, S and H were first determined from samples taken from sides three days after slaughter which had been chilled to $0^{\circ}C$, and then from samples which had been deep frozen for three months. As regards the genetic type, it was observed that the CH and LM breeds caused an increase ($P < .01$) in lightness and yellowness.

Deep-freezing, however, resulted in a reduction ($P < .01$) of lightness (-2%), a_L (-8%), b_L (-7%) and S (-7%). As regards H, an interaction ($P < .05$) with the genetic type of the animal was noted, in that there was a 7% reduction in the CH x GS crosses and an average increase of 3% with other subjects. Finally, "within groups" simple correlations between the variables studied have been reported.

Introduction

Perhaps colour is that characteristic which influences the customer most when buying meat. This physical characteristic does not, however, directly affect the nutritional value of the meat, but it is often used as a reference as it is influenced by genetic type, sex, age, nutritional condition and by the breeding techniques of the animals to be slaughtered, and also by the type of muscle and the length and mode of conservation employed (Romans, Tuma % Tucker, 1965; Rickansrud & Henrickson, 1967; von Böhning et al., 1977).

The aim of this research was to evaluate the effect of deep-freezing, conservation and thawing on the colour of the meat from young bulls belonging to dual-purpose breeds and crosses between these and specialised beef breeds.

Materials and Methods

Sample joints taken from 67 young bulls of the following genetic types were examined :-

8 Italian Brown Swiss (IBS)
12 Piedmont x IBS
11 Limousin x IBS
8 German Simmental (GS)
7 Piedmont x GS
10 Limousin x GS
11 Charolais x GS

The breeding, feeding and slaughering techniques employed were those described by Bonsembiante et al. (1981).

After slaughter, which took place at an average liveweight of 548 kg, the sides were chilled to 0°C for 72 hours, after which the 10th rib sample joint was taken from the right side using the method described by Lanari (1972). After an hour, during which the sample joints were kept at 0°C with one freshly cut surface exposed to the air so that the pigment on the surface could oxygenate to oxmyoglobin, the determinations of colour were carried out according to the method described by MacDougall (1979) on behalf of the Colour Committee, Quality Assessment Group.

The measurements were carried out on the freshly cut surface of the M. Longissimus dorsi with an EEL Reflectance Spectrophotometer,with three replications. The reference system of colour measurement was the 1931 C.I.E. colour space, and the results were reported according to Hunter's system.

The sample joint was dissected in order to determine the incidence of lean, fat and bone, and the section of Longissimus dorsi was sealed in a polythene bag and deep-frozen to -18°C.

After a period of about three months, the samples were thawed out by placing the packages at room temperature. The muscle was then cut across the direction of the muscle fibre and a 2 cm thick slice was kept at 0°C for an hour exposed to the air. The colour analysis was then carried out.

A second slice of M. Longissimus dorsi was examined in order to determine the total content of pigments, measured as haematin by the method of Hornsey (1956), reported by MacDougall (1979).

A third slice was studied in order to evaluate the moisture content, using the freeze-drying process, and the ether extract, using the A.O.A.C. method (1970).

The statistical analysis was carried out according to the experimental design reported by Bonsembiante et al.(1981) and was integrated with simple correlation coefficients between the variables calculated by the cross-product "within-group" matrix. The effect of deep-freezing was evaluated by the "t" test for paired observations.

Results and Discussion

When the chemical composition of the M. Longissimus dorsi was examined, a highly significant difference between the genetic types was noted (Table 1). The Piedmont crosses yielded the highest moisture value and lowest fat content, whereas the opposite was true of the Limousin crosses. The results for the Charolais x Simmental crosses and for the two pure dual-purpose breeds fell between these two.

The low level of ether extract in the muscle of the Piedmont crosses concurs with the low incidence of fatty tissue in the sample joint, and with the modest amount of peri-visceral fatty deposits found in these subjects (Bonsembiante et al., 1981), confirming the scarce fat synthesis ability of this breed.

No significant difference among the various genetic types was noted, however, regarding the total content of pigment measured as haematin. In another paper, (Bittante, 1981) a greater difference was noted between the haematin content of the IBS and GS (167 vs 184 mg/kg), and above all, much more modest values were reported for Charolais and Limousin purebreds and crosses (from 143 to 159 mg/kg).

Taking the physical variables of colour into consideration, it can be seen that these, unlike the pigment content, were significantly influenced by the genetic type of the young bull and also by the technique of freezing employed.

The values of chilled samples showed a high level of lightness for the meat of Charolais and Limousin crosses (Table 2), which corresponded with a high value of yellowness (Table 4),

Table 1 The effect of genetic type on the content of moisture, ether extract and haematin of the Longissimus dorsi samples

	Samples	Moisture %	Ether extract % DM	Haematin mg/kg
1. Italian Brown Swiss (IBS)	8	75.1	5.9	166
2. Piedmont x IBS	12	75.7	3.3	174
3. Limousin x IBS	11	75.3	6.8	184
4. German Simmental (GS)	8	75.1	5.2	173
5. Piedmont x GS	7	76.0	4.1	171
6. Limousin x GS	10	74.9	6.9	161
7. Charolais x GS	11	75.1	6.0	169
OVERALL	67	75.3	5.4	171
Error mean square		0.32	3.61	468
Orthogonal comparisons:				
- dam breed effect (1+2+3vs4+5+6)		NS	NS	NS
- crossbreeding effect (2+3+5+6vs1+4)		NS	NS	NS
- Piedmont vs Limousin (2+5vs3+6)		$P < .01$	$P < .01$	NS
- Charolais effect (7vs1+2+3+4+5+6)		NS	NS	NS
- other comparisons		NS	NS	NS

NS = not significant

Table 2 The effect of genetic type and freezing on the lightness of the Longissimus dorsi samples (L)

	Chilled	Frozen	Difference
1. Italian Brown Swiss (IBS)	38.5	38.1	-0.4
2. Piedmont x IBS	37.0	36.7	-0.3
3. Limousin x IBS	38.0	37.4	-0.6
4. German Simmental (GS)	37.1	37.4	+0.3
5. Piedmont x GS	37.7	36.9	-0.8
6. Limousin x GS	40.2	38.3	-1.9
7. Charolais x GS	39.3	38.6	-0.7
OVERALL	38.2	37.6	-0.6 *
Error mean square	1.41	2.66	1.84
Orthogonal comparisons:			
- dam breed effect (1+2+3vs4+5+6)	NS	NS	NS
- crossbreeding effect (2+3+5+6vs1+4)	NS	NS	NS
- Piedmont vs Limousin (2+5vs3+6)	P<.01	NS	NS
- Charolais effect (7vs1+2+3+4+5+6)	P <.05	NS	NS
- other comparisons	P<.01	NS	NS

NS = not significant

* P<.01

whereas the value of redness was not significantly influenced by the genetic type (Table 3).

Saturation and hue, which provide a representation of colour according to a system of polar coordinates, instead of rectangular coordinates, as is the case with redness and yellowness, were only slightly influenced by genetic type (Tables 5 - 6). Only the Charolais x Simmental crosses stood out from the other subjects in that they yielded a higher value of hue due to the redness/yellowness ratio.

Examining the values of lightness and redness it was observed that an interaction ($P < 0.1$) existed between the dam breed and the effect of crossbreeding, inasmuch as for both of the variables, the use of Piedmont and Limousin sires caused an improvement in the colour of the meat when the dam breed was Simmental, whereas the opposite was true when the dams were Brown Swiss.

The data analysed in this research fully confirm the results of the above-mentioned research (Bittante, 1981) in which the Charolais and the Limousin purebreds or crosses yielded values of lightness and yellowness which were markedly superior to those obtained from Brown Swiss and Simmental. Similarly, the hue results were slightly higher for the Charolais, both for the purebreds and for the crosses.

The only noticeable difference between the two trials concerned the values of redness, which were much higher for the purebred Limousin subjects in the first trial, but not for the Limousin crosses of the present experiment.

Deep-freezing caused an increase of the residual variability of all the determined variables which also corresponded, in the case of lightness, yellowness and hue, to a reduction of the difference existing among the average values of the different genetic types. These two phenomena explain the absence of any significant difference among genetic types concerning the colour of the M. Longissimus dorsi sample which had been deep-frozen and then thawed (Tables 2,3,4,5,6).

Independent of genetic type, deep-freezing caused a 2% reduction ($P < .01$) in lightness, an 8% reduction ($P < .01$) in redness, a 7% reduction ($P < .01$) in yellowness, and finally, a

Table 3 The effect of genetic type and freezing on the redness of the longissimus dorsi samples (a_L)

	Chilled	Frozen	Difference
1. Italian Brown Swiss (IBS)	22.5	21.0	-2.5
2. Piedmont x IBS	25.1	21.9	-3.2
3. Limousin x IBS	24.5	22.1	-2.4
4. German Simmental (GS)	23.4	21.3	-2.1
5. Piedmont x GS	24.3	22.0	-2.3
6. Limousin x GS	24.6	23.9	-0.7
7. Charolais x GS	22.8	23.0	+0.2
OVERALL	24.0	22.1	-1.9 *
Error mean square	6.13	9.14	15.07
Orthogonal comparisons:			
- dam breed effect (1+2+3vs4+5+6)	NS	NS	NS
- crossbreeding effect (2+3+5+6vs1+4)	NS	NS	NS
- Piedmont vs Limousin (2+5vs3+6)	NS	NS	NS
- Charolais effect (7vs1+2+3+4+5+6)	NS	NS	NS
- other comparisons	NS	NS	NS

NS = not significant

* $P<.01$

Table 4 The effect of genetic type and freezing on the yellowness of the Longissimus dorsi samples (b')

	Chilled	Frozen	Difference
1. Italian Brown Swiss (IBS)	18.5	17.0	-1.5
2. Piedmont x IBS	17.7	16.6	-1.1
3. Limousin x IBS	17.9	17.0	-0.9
4. German Simmental (GS)	17.6	17.0	-0.6
5. Piedmont x GS	17.8	16.6	-1.2
6. Limousin x GS	19.1	17.3	-1.8
7. Charolais x GS	18.9	17.3	-1.6
OVERALL	18.2	17.0	-1.2 *
Error mean square	0.48	0.66	0.55
Orthogonal comparisons:			
- dam breed effect (1+2+3vs4+5+6)	NS	NS	NS
- crossbreeding effect (2+3+5+6vs1+4)	NS	NS	NS
- Piedmont vs Limousin (2+5vs3+6)	P <.01	NS	NS
- Charolais effect (7vs1+2+3+4+5+6)	P <.01	NS	NS
- other comparisons	P <.01	NS	NS

NS = not significant

* P<.01

Table 5 The effect of genetic type and freezing on the saturation of the colour of the Longissimus dorsi samples (S)

	Chilled	Frozen	Difference
1. Italian Brown Swiss (IBS)	30.0	27.2	-2.8
2. Piedmont x IBS	30.7	27.5	-3.2
3. Limousin x IBS	30.4	27.9	-2.5
4. German Simmental (GS)	29.3	27.4	-1.9
5. Piedmont x GS	30.1	27.6	-2.5
6. Limousin x GS	31.3	29.6	-1.7
7. Charolais x GS	29.7	28.8	-0.9
OVERALL	30.2	28.0	-2.2 *
Error mean square	3.82	6.37	10.01
Orthogonal comparisons:			
- dam breed effect (1+2+3vs4+5+6)	NS	NS	NS
- crossbreeding effect (2+3+5+6vs1+4)	NS	NS	NS
- Piedmont vs Limousin (2+5vs3+6)	NS	NS	NS
- Charolais effect (7vs1+2+3+4+5+6)	NS	NS	NS
- other comparisons	NS	NS	NS

NS = not significant

* P< .01

Table 6 The effect of genetic type and freezing on the hue of the Longissimus dorsi samples (H)

	Chilled	Frozen	Difference
1. Italian Brown Swiss (IBS)	38.3	39.6	+1.3
2. Piedmont x IBS	35.4	37.5	+2.1
3. Limousin x IBS	36.6	37.9	+1.3
4. German Simmental (GS)	37.1	39.2	+2.1
5. Piedmont x GS	36.4	37.5	+1.1
6. Limousin x GS	37.9	36.0	-1.9
7. Charolais x GS	39.9	37.0	-2.9
OVERALL	37.4	37.8	+0.4
Error mean square	10.05	13.55	21.64
Orthogonal comparisons:			
- dam breed effect (1+2+3vs4+5+6)	NS	NS	NS
- crossbreeding effect (2+3+5+6vs1+4)	NS	NS	NS
- Piedmont vs Limousin (2+5vs3+6)	NS	NS	NS
- Charolais effect (7vs1+2+3+4+5+6)	$P < .05$	NS	$P < .05$
- other comparisons	NE	NS	NS

NS = not significant

7% reduction ($P < .01$) in saturation. The increase of 1% in hue was not considered to be significant. However, it should be noted that, concerning this last-mentioned variable, the only significant interaction observed between the effects of deep-freezing and genetic type, was that whilst meat from Charolais x Simmental crosses was subject to a drop of 7% in hue after deep freezing, that from the other subjects showed an average increase of 3%.

These results were not, however, fully confirmed in a further trial on the effect of deep-freezing on the colour of meat (Bittante & Andrighetto, 1981) carried out on the M. Longissimus dorsi of 16 Polish Friesian young bulls, which yielded only limited variations between the valuations taken after slaughter and after deep-freezing at $-18^{\circ}C$.

To sum up, if the simple correlations between the chemical characteristics and physical variables of colour relative to samples of M. Longissimus dorsi are examined, it can be said that :

a) the contents of moisture and fat in the meat which are closely correlated ($r = -.69$) do not significantly influence colour;

b) the content of haematin is, on the other hand, negatively correlated with the value of lightness and yellowness; it affects, but not significantly, redness, saturation and hue;

c) lightness is closely correlated to redness;

d) saturation and hue are strongly affected by redness and yellowness, from which they are calculated, and in particular by the first, which shows more variability.

While the majority of these results concur with those found in the bibliography, this is not true of the correlation between the pigments and the redness of the meat, which in other research projects, contrary to this one, are positively and significantly correlated (Froning, Daddario & Hartung, 1968; Santoro, Rizzi & Mantovani, 1978).

Table 7 Simple correlation coefficients between colour index and moisture, ether extract and haematin contents of the Longissimus dorsi samples

	L	a_L	b_L	S	H
L.D. contents :					
- moisture	+.04	-.14	-.03	-.13	+.14
- ether extract	+.07	+.22	+.13	+.24	-.17
- haematin	-.50 **	-.15	-.43 **	-.24	-.02
L.D. colour:					
- lightness (L)		+.06	+.89 **	+.24	+.28*
- redness (a_L)	+.06		+.13	+.98 **	-.93 **
- yellowness (b_L)	+.89 **	+.13		+.33 **	+.24
- saturation (S)	+.24	+.98 **	+.33 **		-.83 **
- hue (H)	+.28 *	-.93 **	+.24	-.83 **	

* $P < .05$

** $P < .01$

Conclusion

This research has shown that compared to purebred Brown Swiss and Simmental and to Piedmont crosses, crossbreeding with Charolais and Limousin bulls does improve the colour of the meat.

Another important result obtained is that deep-freezing, cold storage, and thawing of meat cause a deterioration of colour, in that the values of lightness, yellowness, redness and saturation are reduced. It also causes an increase in the residual variance and a reduction of the differences which can be attributed to genetic type. However, further research is recommneded in order to evaluate the effective correspondence existing between the colour and the nutritive value of the meat, and to investigate more thoroughly the variation of these physical characteristics.

References

A.O.A.C. (1970). Official Methods of Analysis. A.O.A.C. Washington, D.C.

Bittante, G. 1981. Caratteristiche delle carni di vitelloni in otto differenti tipi genetici. In press.

Bittante G. and Andrighetto, I. 1981. Unpublished data.

von Böhning, V., Hildebrandt, E., Weiher, O. and Neumann, W. 1977. Der Einfluss des Genotyps und der Mastendmasse auf den Protein und Fettgehalt sowie die Fleischfarbe des M. longissimus dorsi und des Caput longum des M. triceps brachii bei Jungbullen. Wissenschaftliche Zeitschrift der Wilhelm Pieck Universitat, Rostock, Mathematisch - Naturwissenschaftliche Reihe, 26, 35.

Bonsembiante, M., Bittante, G., Cesselli, P. and Andrighetto, I. 1981. Fattening performance and slaughter traits of purebred and crossed young bulls. EEC Seminar on Beef Production from the Dairy Herd , Dublin.

Froning, G.W., Daddario, J. and Hartung, T.E. 1968. Colour and Myoglobin concentration in turkey meat as affected by age, sex and strain. Poultry Sci. 47, 1827.

Hornsey, H.C. 1956. J. Sci. Fd. Agric. 1, 534 - cited by MacDougall, D.B. (1979).

Lanari, D. 1972. Osservazioni sulla scelta dei parametri per la valutazione delle carcasse di vitelloni e calcolo dei coefficienti di accrescimento dei tessuti muscolare, adiposo ed osseo. Riv. Zoot. 45, 77.

MacDougall, D.B. 1979. Recommended procedures for use in the measurement of meat colour. The Future of Beef Production in the European Community, Martinus Nijhoff, The Hague.

Rickansrud, D.A. and Henrickson, R.L. 1967. Total pigments and myoglobin concentration in four bovine muscles. J. Food Sci. 32, 57.

Romans, J.R., Tuma, H.J. and Tucker, W.L. 1965. Influence of carcass maturity and marbling on the physical and chemical characteristics of beef. II Muscle pigments and colour. J. Anim. Sci. 24, 686.

Santoro, P., Rizzi, and Mantovani, C. 1978. Colore e contenuto di mioglobina in carni di tacchini macellati a diverse età. Zoot. et Nutr. Anim. 4, 147.

EARLY CALVING OF CROSSBRED HEIFERS. 1. COMPARISON BETWEEN ANIMALS SLAUGHTERED JUST AFTER AND SEVEN MONTHS AFTER FIRST CALVING

A. Romita, S. Gigli, A. Borghese,
A. Di Giacomo, M. Mormile, C. Esposito

Istituto Sperimentale per la Zootecnia, Via Salaria 31,
00016 Monterotondo Scale, Roma, Italy.

Summary

A trial was carried out on 179 heifers of five genetic types (Chianina, Limousin , Marchigiana, Piemontese, Charolais) x Friesian to study the problems related to an early calving. The carcass value of the animals slaughtered ten days after first calving, or after weaning of their calf was also studied.

With the exception of the Chianina x Friesian cross, the incidence of stillbirths and the total mortality of calves was very high.

The carcass value of the animals slaughtered after calf weaning seemed to be better. Piemontese x Friesian seemed to have better meat.

Introduction

Most calves fattened in Italy come from foreign countries. The price of the imported young animals is continuously increasing so that the income for the breeders is becoming lower and lower. In this situation it is necessary to increase the total number of homebred calves.

As it is difficult to increase the number of cattle, one way of producing more calves is to utilise all the females not kept as replacements (Romita, 1973; Owen, 1973) but that are normally slaughtered at a young age. If these animals produce one or more offspring before slaughter, it will serve a dual purpose: more calves and higher slaughter weight from the mothers.

To optimise this breeding strategy it is necessary to reduce the age at first calving in order to supply carcasses and meat of good quality, to provide animals with the cheapest food, to utilise crossbred animals, and to find the genetic types which are most useful, and also the most profitable age for slaughter.

Since 1969 we have been studying the problems concerning once bred heifers and early calving (Romita, 1973, 1975). Relevant studies have been carried out also by Bonsembiante et al.(1977, 1979a,b). A complete review is reported by Morris (1980a,b).

The importance of this item was also stated by the CEC Beef Committee which retained in its common programme some projects and organised its first seminar on this topic (1975). In this paper we report on the first results of an EEC programme on early calving.

Material and Methods

This experiment involved 179 animals subdivided into five genetic types :-

34 Chianina x Frisona	CxF		39 Piemontese x Frisona	PxF
33 Limousin x Frisona	LxF		38 Charolais x Frisona	SxF
35 Marchigiana x Frisona	HxF			

The animals were reared for 6-7 months indoors and afterwards in the open air and fed according to the Scandinavian method. They received milk substitute and concentrates up to 140 kg and thereafter hay, grass and maize silage.

The heifers were mated for the first time from 12 to 15 months by many Piedmont bulls.

The cows were slaughtered according to the following scheme :

Group A 25% 10 days after first calving
Group B 25% 8 months after first calving
Group C 25% 10 days after second calving
Group D 25% 8 months after second calving

The cows of Groups B and D suckled their calves for 7 months. At slaughter, body and carcass measurements, carcass and five quarter weights were taken. Eight days after slaughter, fatness and conformation were evaluated according to the method of De Boer et al. (1974). Seven animals of each genetic type and for each Group were completely dissected according to the method of Williams and Bergström (1976).

With regard to the muscles, reported in Table 7a, colour was determined using an E.E.L. apparatus, and tenderness with a

W.B.S. (Instron 1140). In this paper we will report only some results concerning Groups A and B, leaving to future papers : regions, muscles, fat and bone distribution in carcass, tissue composition of single region, chemical and physical determination of meat and fat, and taste panel results.

Results

1. Body weight and feed consumption

As is shown in Figure 1, the body weight of LxF and SxF was very similar and generally higher than that of the other genetic types at all the ages considered; lowest weights were shown by the animals LxF. Weights increased up to calving and at conception varied from 340 to 370 kg..

Daily gain varied from 641 to 696 g, close to the 700 g scheduled, when the experiment was set up, which seemed the optimum to obtain good results with early calving (Neumann *et al.*, 1974; Petit, 1975). Daily gain, from the start to post partum, varied from 577 g (SxF) to 523 g (LxF): this last genetic type differed significantly from the others. The daily average S.F.U. consumed was 4.5 and 7.0 up to calving, and up to weaning, respectively. The 7 S.F.U. consumed by the cows included the food consumed by their calves.

2. Calving and weaning performance

As shown in Table 1, calving occurred between 25 and 26 months of age. LxF and PxF had more difficulty at calving. The mortality at calving was low, except for PxCF; the stillbirth rate of PxPF was dramatic and the low viability of these animals was also evident from the high mortality which occurred in later stages. The PxSF mortality occurred within the first week. The total mortality of the LxF was also high, while among the other genetic types it was nearly 20%. The mortality after calving was high, probably because many of the calves were born in winter: therefore a reduction could be attained if the cows calved indoors, or in a better season.

As expected, the weights of male calves are higher than those of females, particularly in PxPF calves. The PxCF were the heaviest and therefore it seemed that the dam and calf shape

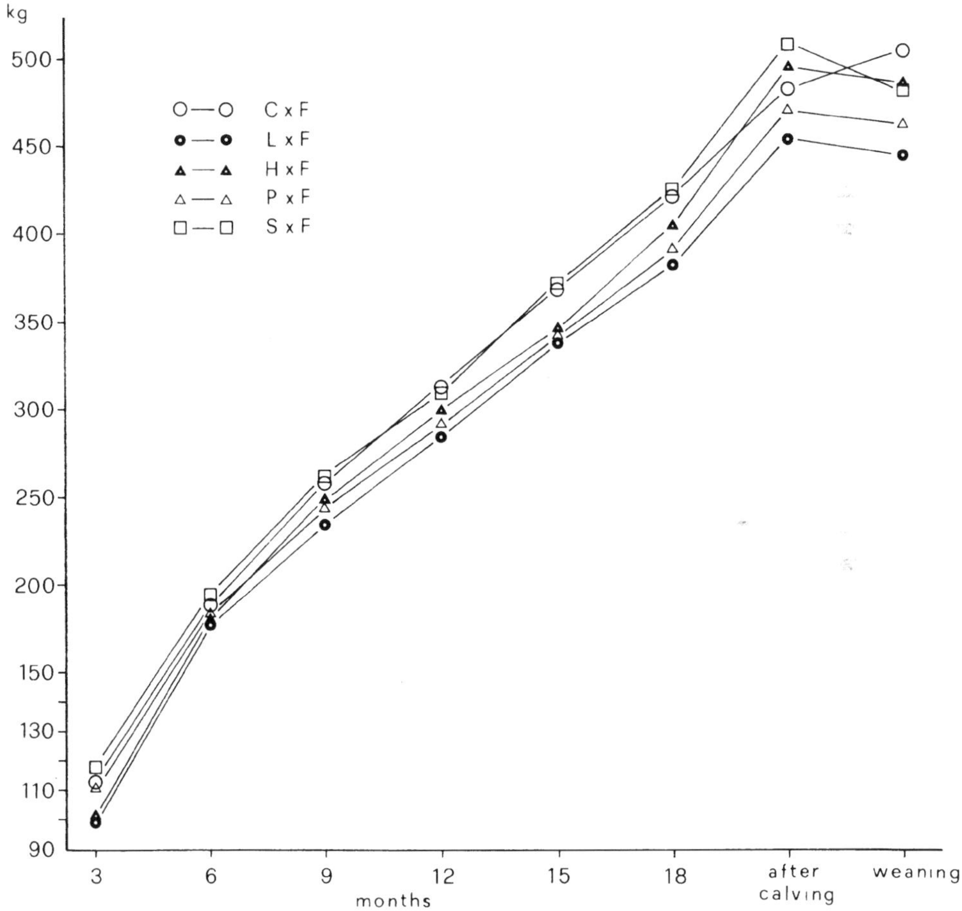

Fig 1 – Daily gain trend

Table 1 Calving and weaning performance

	Genetic type of cows				
	C x F	L x F	H x F	P x F	S x F
Age at calving (months)	25.4	26.0	26.0	24.8	25.4
Calving difficulties*	1.58	2.34	1.67	2.24	1.80
Calf mortality					
Total number	7	10	6	19	7
% born calves	22.4	34.4	19.9	49.9	19.5
% stillborn	3.2	10.3	10.0	28.9	16.7
% dead in 7 days	6.4	6.9	3.3	10.5	2.8
% dead after 7 days	12.8	17.2	6.6	10.5	0.0
Cow's weight after calving (kg)	483.3a	450.5b	497.4a	471.3ab	491.1a
Calf weight (kg)					
Males	43.4	40.7	38.4	42.1	39.6
Females	38.0	34.9	37.1	34.2	35.5
All	41.7	38.3	37.7	38.9	37.2
Cow's weight at weaning of calf (kg)	500.6a	445.7c	486.6ab	462.2bc	481.9ab
Calf weight at weaning (kg)					
Males	222.0	211.1	207.9	202.5	215.9
Females	208.0	206.1	203.5	186.9	209.7
All	217.0	208.0	205.3	196.3	212.8

* 1 = spontaneous 5 = caesarian
Different letters in the same line mean significant difference among genetic types

besides the weight of calves, affected the mortality at calving. (Smith et al. 1976). Little variation was found between the cow weights after calving and after weaning, as found by Boucque et al. (1980). The weights of the PxCF and PxSF calves at weaning were higher (about 20 kg with respect to the PxPF) but the differences were not significant. Daily gain varied from 797 g (PxCF) to 739 g (PxPF).

3. Slaughtering performance

In Tables 2 and 3, body and carcass measurements (Fig. 2) at slaughter are reported.

The CxF measurements are generally higher and often significant in particular for the heights and lengths. There were no significant differences between groups, except that the PxF animals of Group A were significantly lighter (Table 4). Net dressing percentages were not different either between genetic types or Groups. Only in the PxF animals did the proportion of forequarter to hindquarter change between the Groups (Table 4).

In Table 5 the percentage of some components of the fifth quarter are reported. Significant differences among genetic types were found only for the empty intestines (Group A) and the skin (Group B). Between Groups there was a trend for the percent of components of the fifth quarter to increase, particularly with regard to the fat covering and the stomach. The PxF animals showed the highest quantity of dissected meat and the lowest percentage of fat, but the differences were not significant (Table 6). The CxF animals tended to have more bone (significant value in Group A). The animals after weaning had generally less meat and bone, due to the increasing amount of fat. The significant differences are reported in Table 6. The increase of fat with age was not evident from the fatness scores (Table 6). The quantity of fat in the carcasses of Group A animals must be considered insufficient.

The animals showed poor conformation and the differences between Groups and genetic types were limited.

The colour (brightness and purity) measurements of ten muscles are reported in Tables 7a and b. The significant

Table 2 Body measurements (cm)

Groups	Height at withers A	Height at withers B	Height at pelvis A	Height at pelvis B	Body length A	Body length B	Length of rump A	Length of rump B	Width at pelvis A	Width at pelvis B
Genetic types										
C x F	135.7a	136.0a	143.6a	139.2a	154.3	157.5a	50.0	49.8a	49.4	47.5
L x F	124.7c	124.8c	127.8c	128.7c	147.0	144.2b	47.3	47.3b	46.5	45.9
H x F	131.4ab	130.9b	136.0b	134.6b	150.7	153.9a	48.6	49.0ab	49.1	47.0
P x F	122.6c**	129.7b	127.2c*	133.5b	143.4**	154.3a	46.8**	49.1ab	45.6	46.6
S x F	127.2bc	126.7bc	133.2b	131.5bc	149.4	146.4b	46.9	47.6b	46.2	46.4

Groups	Depth of chest A	Depth of chest B	Width of chest A	Width of chest B	Chest girth A	Chest girth B	Spiral round A	Spiral round B
Genetic types								
C x F	69.0ab	67.5a	40.7	39.4	184.4ab	185.7	193.4	198.6
L x F	64.5bc	63.5b	40.2	39.6	177.8bc	178.4	189.3	191.0
H x F	65.6ab	67.0a	39.9	40.6	180.1bc	186.4	194.5	199.0
P x F	61.0c**	65.4ab	38.0	40.8	172.0c**	183.7	181.8***	197.5
S x F	66.5ab	66.0ab	42.3	41.6	186.8ab	184.6	199.6	194.9

* means significant difference between groups

different letters in the same column mean significant differences among genetic types

Table 3 Carcass measurements (cm)

Groups	Carcass length A	Carcass length B	Round length A	Round length B	Maximum length of round A	Maximum length of round B	Minimum length of round A	Minimum length of round B	Depth of chest A	Depth of chest B
Genetic types										
C x F	136.0a	136.3a	74.5a	70.8a	25.3	21.1	25.2	24.9	47.5a	49.0a
L x F	130.3ab	129.6b	67.4b	66.6b	24.9	25.1	23.5	24.1	45.6ab	44.5b
H x F	132.4ab	133.4ab	71.2ab	68.0b	25.2	25.6	25.1	24.9	45.0ab	46.2b
P x F	127.6b**	134.5a	67.6b	67.3b	24.3	25.3	24.1	24.6	41.4b*	45.8b
S x F	133.8ab	133.0ab	68.4b	67.4b	24.9	25.5	24.6	24.8	45.7ab	45.3b

Table 4 Slaughter data

Groups	Net live wt. kg		Net dressing percentage		Percentage on right side							
					Forequarter		Hindquarter		Kidney fat		Pelvic fat	
	A	B	A	B	A	B	A	B	A	B	A	B
Genetic types												
C x F	408.2a	404.8	59.2	59.2	51.7	52.1	48.3	47.9	0.96	1.30	0.14	0.20
L x F	371.8a	378.4	60.0	60.4	52.4	52.2	47.6	47.8	1.40	1.33	0.15	0.17
H x F	379.7a	412.2	59.6	60.8	51.8	52.9	48.2	47.1	1.21	1.73	0.14	0.20
P x F	327.6b***	400.9	60.6	61.0	51.6	52.9	48.4	47.1	0.93	1.10	0.15	0.13
S x F	406.6a	389.7	59.1	60.1	52.6	52.8	47.4	47.2	1.53	1.23	0.15	0.15

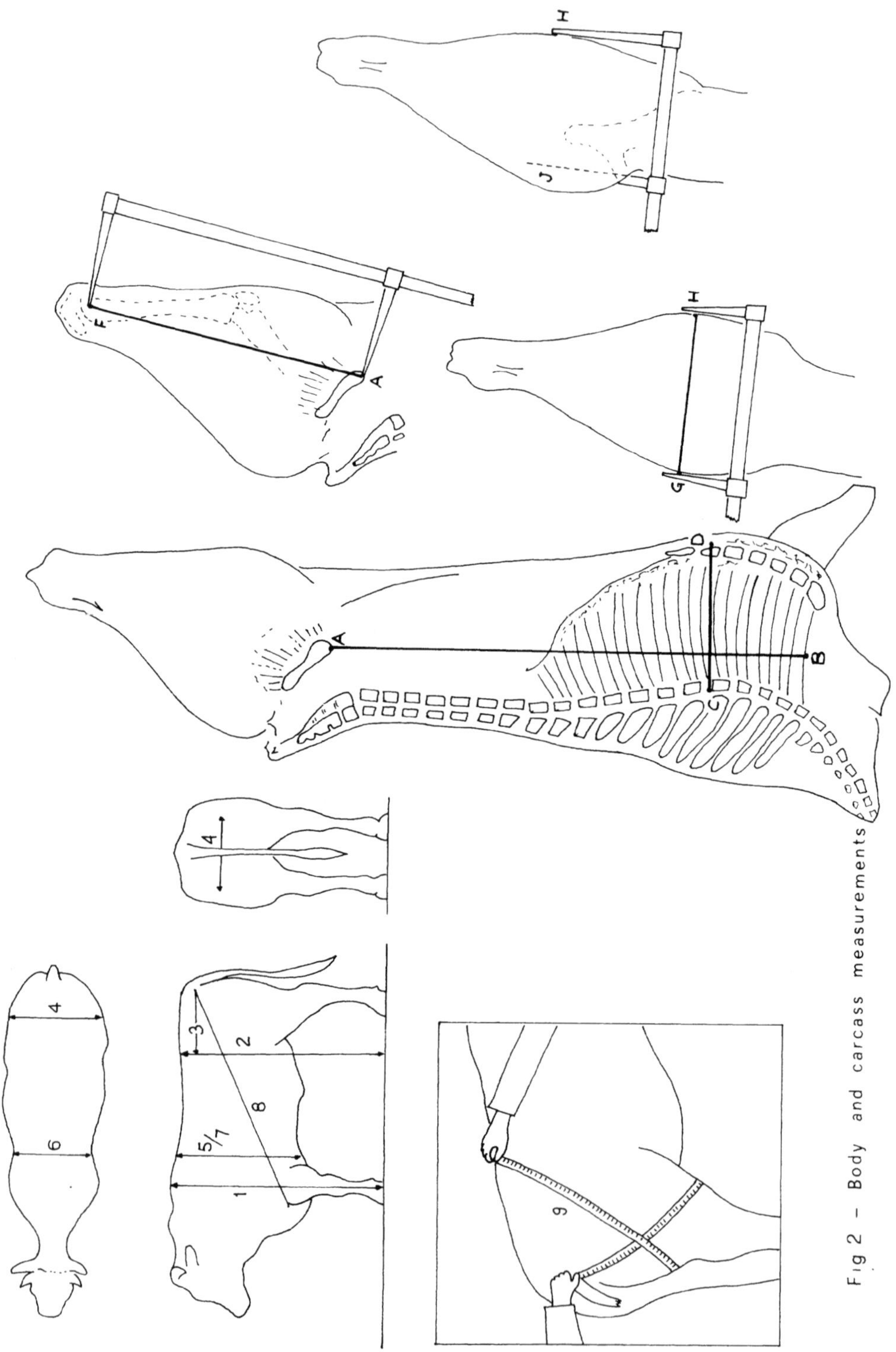

Fig 2 – Body and carcass measurements

Table 5 Fifth quarter; percentage on net liveweight

Groups	Empty stomach A	Empty stomach B	Empty intestine A	Empty intestine B	Stomach fat A	Stomach fat B	Intestinal fat A	Intestinal fat B	Skin A	Skin B
Genetic types										
C x F	3.94	4.40	2.67b**	3.20	0.48	0.75	0.40	0.43	8.51	8.63a
L x F	3.78***	4.65	2.94ab*	3.33	0.68	0.78	0.57	0.44	8.30	8.12a
H x F	4.23	4.24	2.87ab	3.22	0.72	0.98	0.50	0.57	8.49	8.54a
P x F	4.56	4.50	3.43a	3.21	0.45*	0.80	0.33	0.43	8.47	7.96a
S x F	4.31	4.48	3.28a	3.24	0.74	0.88	0.42	0.56	7.71	7.81b

Groups	Head A	Head B	Liver A	Liver B	Heart A	Heart B	Lungs A	Lungs B	Spleen A	Spleen B
Genetic types										
C x F	3.28	3.21	1.44	1.47	0.45	0.45	1.10	1.13	0.19	0.20
L x F	3.25	3.21	1.34*	1.50	0.42	0.45	1.10	1.04	0.17	0.19
H x F	3.21	3.24	1.36*	1.54	0.46	0.49	1.04	1.08	0.19	0.20
P x F	3.41	3.22	1.54	1.49	0.45	0.43	1.14	1.00	0.20*	0.17
S x F	3.23	3.27	1.41	1.51	0.45	0.46	1.15	1.01	0.18	0.20

Table 6 Carcass composition and evaluation

Groups	Meat %		Bone %		Subcutaneous fat %		Intermuscular fat %		Total fat %	
	A	B	A	B	A	B	A	B	A	B
Genetic types										
C x F	68.45	67.67	22.88a *	20.48	2.84 *	3.90	4.70 **	6.67	7.54 *	10.57
L x F	69.29	69.06	20.19b	19.82	3.53	3.67	5.77	6.16	9.30	9.83
H x F	69.41	68.09	21.02ab	19.58	3.17 *	4.45	5.19 *	6.52	8.36 *	10.97
P x F	71.58	70.36	20.72ab	19.52	2.57	3.40	4.04	5.48	6.61	8.88
S x F	68.18	68.87	20.89ab	20.57	3.43	3.23	6.38	6.18	9.81	9.41

Groups	Other tissues %		Conformation		Fatness	
	A	B	A	B	A	B
Genetic types						
C x F	1.13	1.28	2	2	1+	1+
L x F	1.22	1.29	2+	2+	1+	1+
H x F	1.21	1.36	2+	2+	1+	2-
P x F	1.09	1.24	2 *	2+	1	1+
S x F	1.12	1.15	2	2	1+	1+

Table 7a Colour measurements - brightness

	2		4		5		7		9	
Groups	A	B	A	B	A	B	A	B	A	B
Genetic types										
C x F	11.48	13.54	12.42	13.13	13.15	12.16	13.52	15.50	12.20	12.40
L x F	10.63 *	12.82	11.73 *	13.78	12.63	13.28	12.30 *	16.42	10.83 **	15.02
H x F	12.27	11.80	12.37	12.10	12.22	12.42	13.92	14.65	12.25	12.13
P x F	12.97	13.56	13.22	13.28	13.62	14.00	14.60	15.90	12.32	13.06
S x F	12.40	13.23	11.64	12.78	13.58	12.70	14.68	14.04	11.78	12.78

	11		12		13		14		15	
Groups	A	B	A	B	A	B	A	B	A	B
Genetic types										
C x F	13.68	14.37	15.35b *	18.64	12.53	12.50b	12.58	13.81	11.18 *	13.78
L x F	12.38	14.60	15.10b *	17.82	12.70	14.83ab	11.47 **	14.07	12.76	13.50
H x F	13.13	13.63	14.80b *	16.12	12.48	12.75ab	12.88	13.03	11.84	13.33
P x F	13.58 *	16.26	19.28a	19.43	13.12	15.14a	12.78	14.47	13.98	15.16
S x F	13.28	14.56	17.88a	16.06	12.76	12.93ab	12.56	13.84	11.88	13.51

2) Pectoralis profundus
4) Triceps brachii
5) Supraspinatus
7) Rectus femoris
9) Gluteus medius
11) Gluteobiceps
12) Semitendinosus
13) Semimembranosus
14) Longissimus dorsi
15) Psoas major

Table 7b Colour measurements - purity

	2		4		5		7		9	
Groups	A	B	A	B	A	B	A	B	A	B
Genetic types										
C x F	33.23	38.52	37.07	35.29	30.93	38.14	36.46	35.37	32.02	37.52
L x F	37.15	37.29	41.87 *	31.70	36.12	36.51	37.85	31.56	40.26	31.39
H x F	37.85	33.38	36.97	32.68	35.58	30.42	42.23	35.25	37.33	32.96
P x F	38.09	41.26	36.48	36.17	41.01	33.80	33.28	32.65	36.28	36.08
S x F	33.11	39.21	38.97	34.29	33.75	34.28	29.01	33.27	33.86	35.77

	11		12		13		14		15	
Groups	A	B	A	B	A	B	A	B	A	B
Genetic types										
C x F	32.81	32.05	34.40	31.41	37.78	34.64	32.64	27.54	40.42	38.02
L x F	37.07	31.08	33.81	32.70	37.10	38.34	38.54	36.33	32.57	33.73
H x F	35.87	29.46	38.58 *	31.45	37.11	35.05	36.96	31.91	41.09	34.50
P x F	35.18	33.92	30.28	30.32	35.53	35.31	41.22	35.56	35.95	35.66
S x F	30.53	25.92	32.19	31.80	34.58	33.45	36.74	31.40	35.77	36.76

2) Pectoralis profundus
4) Triceps brachii
5) Supraspinatus
7) Rectus femoris
9) Gluteus medius
11) Gluteobiceps
12) Semitendinosus
13) Semimembranosus
14) Longissimus dorsi
15) Psoas major

Table 8 W.B.S. values (kg)

	2		4		5		7		9	
Groups	A	B	A	B	A	B	A	B	A	B
Genetic types										
C x F	15.05	14.73	13.45	12.07	17.68	19.20	10.41	9.38	11.22a *	9.90
L x F	13.87	14.20	11.38	10.28	18.00	16.90	9.67	8.06	9.58ab	10.07
H x F	14.59	14.96	11.43	10.95	18.19	18.97	9.04	8.51	9.63ab	9.96
P x F	14.44	14.32	11.22	12.06	15.64	17.77	9.32 **	7.69	9.21b	9.02
S x F	15.19	15.15	11.77	12.38	18.14	19.18	8.30	9.42	10.04ab	10.14

	11		12		13		14		15	
Groups	A	B	A	B	A	B	A	B	A	B
Genetic types										
C x F	20.63	20.29	19.78	19.35	18.81	18.62a	10.01a	9.03	7.48	6.85
L x F	20.32	20.73ab	19.38	18.59	18.54	17.75ab	7.92b	8.64	6.48	6.12
H x F	20.05 *	22.23a	18.62	19.18	18.47	18.09ab	8.52ab	8.23	6.56	6.61
P x F	17.97	19.68b	17.72	17.17	16.83	16.13b	7.86b	7.53	6.07	5.51
S x F	20.48	18.97b	19.43	18.46	18.35	17.16ab	9.49ab	8.37	6.92	6.03

2) Pectoralis profundus
4) Triceps brachii
5) Supraspinatus
7) Rectus femoris
9) Gluteus medius
11) Gluteobiceps
12) Semitendinosus
13) Semimembranosus
14) Longissimus dorsi
15) Psoas major

differences were very few, but the general trend was for the semitendinosus muscle to have the lightest colour; most of the PxF muscles had a lighter colour. The animals of Group A had darker muscles, probably because of the stress of calving and the poorer quantity of intramuscular fat. The change in brightness for the muscles of LxF animals that became lighter in Group B was evident, and often significant. As is shown in Table 8, the ten muscles could be subdivided into four tenderness categories, as follows :-

Classification	W.B.S. value	Muscle
Very tender	7.5	15
Tender	7.5 - 13.5	4-7-9-14
Slightly tender	13.5 - 19.5	2-5-13
Tough	19.5	11-12

The muscles of PxF were generally more tender and some differences were significant; there was no definite trend between groups.

Conclusions

As the differences among genetic types up to the age considered, for daily gain, carcass composition and meat characteristics were limited, stillbirth rate and calf mortality are the most important problems in once-bred heifers. It appears that the Piedmont bulls give many problems at calving, especially with Piedmont crossbred females and should only be used with CxF; this is confirmed by a new experiment which we are currently carrying out on 40 CxF animals, where the stillbirth rate was 8%. Nevertheless, as the Piedmont bulls produce calves with very good conformation and good quality meat, it could be profitable to use these bulls on older cows and to find lines of bulls that produce easy calving. In fact, in another trial, using Limousin bulls, which are normally associated with easy calving, once-bred heifers of different genetic types had stillbirth rates from 10 to 27% (Romita et al., 1981).

To choose a profitable genetic type for once-bred heifers it would be interesting to know also the performance of their offspring which are actually being fattened in our Institute.

The differences between the animals slaughtered just after first calving or after weaning are not evident, but carcass and meat characteristics seem better in the latter group. Therefore, the two systems are comparable and one can choose between them depending on economical and environmental factors.

References

Bonsembiante, M., Chiericato, G.M. 1977. Allevamento della manza in ambiente confinato per la produzione del vitello: effetto di differenti fonti di integrazione azotata della razione. Zoot. Nutr. Anim. 3, 41-52.

Bonsembiante, M., Lanari, D., Susmel, P. and Andrighetto, I. 1979a. Results of a confinement trial with once bred German Simmental heifers. In "The future of beef production in the European Community". Ed. J.C. Bowman, and P. Susmel. Abano Terme, 13-17 novembre.

Bonsembiante, M., Rioni-Volpato, M., Bittante, G. and Guidetti, G. 1979b. Allevamento confinato di manze meticce per la produzione del vitello: effetto della combinazione razziale materna e cause di distocia. Zoot. Nutr. Anim. 5, 613-633.

Boucque, Ch.V., Fiems, L.O., Cottyn, B.G., Buysse, F.X. 1980. Beef production with maiden and once-calved heifers. Livestock Prod. Sci. 7, 121-133.

De Boer, H., Dumont, B.L., Pomeroy, R.W. and Weniger, J.H. 1974. Manual on EAAP reference methods for the assessment of carcass characteristics in cattle. Livestock Prod. Sci. 1, 151-164.

Morris, C.A. 1980a. A review of relationships between aspects of reproduction in beef heifers and their lifetime production. 1. Associations with fertility in the first joining season and with age at first joining. Anim. Breed. Abstr. 48, 655-676.

Morris, C.A. 1980b. A review of relationships between aspects of reproduction in beef heifers and their lifetime production. 2. Associations with relative calving date and with dystocia. Anim. Breed. Abstr. 48, 753-767.

Neumann, W., Zupp, W., Sperlich, W., Bietz, G. , Kaufmann, O. and Schenke, M. 1974. Ergebnisse aus Untersuchungen zur Vornutzung von Mastfärsen. 1. Mitteilung: Aufzucht und Fruchtbarkeit, Trächtigkeit und Kalbung, Saugereit und Absetzen. Arch. Tierzucht. 17, 227-236.

Owen, J.B. 1973. The maiden female - a means of increasing meat production. Proc. Symp. on the once - bred heifers and gilts. 11 Oct. 1972, Aberdeen, 71 pp.

Petit, M. 1975. Vêlage précoce dans les troupeaux de vaches allaitantes. Bull. Tech. CRZV. No. 22, p.5.

Romita, A. 1973. Parto precoce in soggetti di razza Frisona Ann. 1st Sper. Zootec. 6, 207-222.

Romita, A. 1975. Early calving. Research results from Italy. In "The early calving of heifers and its impact on beef production." Ed. J.c. Tayler, Copenhagen, 4-6 giugno.

Romita, A., Borghese, A., Gigli, S., Mazziotti di Celso, P., Chimini G. 1981. Effetto del parto in età precoce sulla

produzione della carne in meticce Frisone. Convegno ASPA, Catania.
Romita, A., Gigli, S. and Borghese, A. 1981. Influenza dell' eta al primo parto sulla carriera produttiva di vacche Frisone. Convegno ASPA. Catania.
Smith, G.M., Laster, D.B. and Gregory, K.E. 1976. Characterization of biological types of cattle. 1. Dystocia and preweaning growth. J. Anim. Sci. 43, 27-36.
Williams, D.R and Bergstrom, P.L. 1976. Anatomical jointing, tissue separation and weight recording proposed as the EEC standard method for beef. EUR 5720.

DISCUSSION ON DR. A. ROMITA'S PAPER

Chairman: Dr. R. Thiessen

More O'Ferrall: In view of the extremely high stillbirth and calf mortality rates when using Piedmont bulls, would it not be better to use a different breed of sire ?

Romita: Yes, and we are looking for example to such breeds as the Limousin in a new series of trials.

More O'Ferrall: Did the nutrition of the heifers prior to calving have an effect on calf mortality ?

Romita: Yes.

Langholz: You have classified tenderness using selected muscle groups. Did you analyse more muscle groups, and what was the principal of sampling ?

Romita: The approach we used was to analyse 10 of what we considered to be the most representative of the carcass muscles.

Menissier: In this experiment you produced a threequarter Piedmont animal. In our experience this will increase the frequency of double muscling problems, and consequently, the level of calving difficulties.

Romita: Yes, I would agree with that.

Bittante: Referring to the problem of controlling birth weight in early calving heifers, we have observed average birth weights to be between 40 - 46 kg amongst heifers of 10 different crosses when sired by Limousin bulls. These results are very close to those obtained by Romita, even though he used Piedmont crosses.

The results we have obtained from several trials indicate that the birth weight of calves out of early-calving heifers, can be effectively controlled with a proper feeding programme. We found that good reproductive performance was obtained using a low level of nutrition, particularly during the last third of pregnancy. Another result was that at a similar level of nutrition, better performance was achieved with a diet of ad libitum roughage, rather than a restricted concentrate diet. Finally, there was evidence that the control of calf birth weight through nutritional means actually reduced the difference in birth weight between sexes, within breeds. We are pursuing this result further in a study involving 330 Charolais cross heifers mated with Charolais or Limousin bulls and maintained on various rations.

SELECTION OF BEEF BULLS FOR TERMINAL CROSSING IN FRANCE

J.L. Foulley and F. Menissier

Station de Génétique quantitative et appliquée,
Centre national de Recherches Zootechniques, I.N.R.A.
78350 Jouy-en-Josas, France.

Summary

Some important topics of the present time in the selection of beef bulls for corssing on dairy populations are discussed especially in relation to French conditions. Performance-test procedures now applied in this country for selecting growth efficiency and potential are presented as a basis for discussion of related problems. According to the importance of birth conditions in the beef-dairy crossing practice, the influence of the main variable factors are reviewed. In that case, sire evaluation on birth weight should be preferred to that on frequency of dystocia. Different selection alternatives for limiting dystocia and improving postnatal growth are examined taking into account age of females crossed and genetic merit of beef breeds chosen for terminal crossing. Finally, the number and types of sire lines to be selected are discussed considering the diversity of selection goals, breeds used or specialised strains developed for that purpose as well as present genetic and economic constraints against too high a degree of specialisation. In the crisis context now affecting selection units and beef dairy crossing in France, a national, even perhaps European plan is necessary to keep an efficient selection of terminal sire lines.

Introduction

Considering the whole production (milk and beef), Cunningham (1974.b) has shown the favourable effect of beef crossing in a dairy population. Theoretically, the total benefit per insemination is proportional to the rate of beef crossing and to the extra beef value of a crossbred over a purebred calf. Therefore, the economic advantage of beef crossing could be very variable according to the beef breeds used, dairy genotypes concerned, culling strategy of cows and tradition of beef crossing practice. However, beef crossing appears especially attractive as it does not conflict with the efficiency of dairy selection. On the contrary, the higher the beef crossing rate

Table 1: Some statistics about AI and beef selection programmes in France over 1970-78.

Year	No of first beef AI (a)	Performance test: no recorded	Performance test: p.100 selected	Progeny-test: no recorded	Progeny-test: p.100 selected
70	2 614 014	324	49	199	42
75	2 053 439	167	57	95	49
78	1 860 480	181	56	93	46
70 - 78	-753 534	-142		-104	
	(-4.15% per year)	(-44%)		(-52%)	

(a) Blond d'Aquitaine, Charolais, Limousin and crossbred bulls

and/or the cow turn-over, the greater the probability of obtaining a dairy heifer from one insemination. This implies a better profitability for dairy selection programmes, the genetic merit being transmitted to the population by fewer inseminations (Cunningham and McClintock, 1974). Summing up, if one wants definitively to apply the most efficient selection for milk, in a dairy population without damaging meat production, one should practice the maximum beef crossing rate the population can accept.

Even if the situation in some countries has been relatively closer to this theoretical pattern such as in Ireland and to a lesser extent in Great Britain, the trend observed practically in some other countries is quite different : either most of the meat production comes out of dairy or dual purpose populations conducted in purebreeding, or specialised beef genotypes are developing in purebred suckling cows, the part of crossbreeding remaining low and/or being just an intermediate stage in grading up and purebreeding. France occupies a middle position between these two extremes. Up to recently (mid 1960's), the rate of beef crossing was considerable (20 to 25p.100). It mainly took place on small mixed farms, located in the centre and south-west, raising milk cows of dual purpose type or suckling cows of hardy and beef crossed breeds producing veal calves of high grade. Accordingly, the 1.7 million inseminations in beef crossing have led the AI units to set up progressively after 1958 an integrated selection scheme in three stages i.e. planned matings, performance testing and progeny-testing, the efficiency of which was clearly demonstrated (Frebling et al., 1971; Gaillard et al., 1974). The disappearance of many small farms, the specialisation of dairy cattle, the great demand for replacement heifers and the decline of the traditional veal calf production were followed after 1970 with a strong recession of beef AI activities (Table 1). Since 1975, this has resulted in a new direction of selection programmes. It may then provide us the opportunity to discuss the objectives, criteria, tools and more general strategy of selection that must be worked out for the choice of beef bulls to be used in crossing on dairy populations. Many aspects related to this topic have been already presented (viz Cartwright, 1970; Cunningham, 1974.a; Lindhe, 1976; Foulley, 1976; Foulley and Menissier, 1979). We would therefore, deal only with some rather important problems of the present time, particularly in the context of French conditions.

1. Selection for Growth Potential

Selecting within terminal sire breeds could be a very economicially efficient way to improve growth characteristics of dairy-beef crossbred progeny as shown by several authors (Mocquot, 1972; Cunningham, 1974. b), especially in comparison with similar selection carried out directly within the dairy breed (Hill, 1971). As reported by Mocquot and Foulley (1973), the relative merit of perforamnce and progeny-test for the above purpose can be appreciated on the basis of a cost-benefit analysis of selection activities. The results of their analysis showed that, in a situation of progeny-testing on veal production, the main factors involved are heritabilities of traits measured in performance and progeny-test, their genetic correlation as well as the proportions of veal and yearling crossbred progeny marketed. In fact, after a short period (1974-78) of progeny testing AI beef bulls on their Friesian crossbred progeny in station for young bull or veal production, progeny testing procedures are moving back to a more simplified field recording systems (birth data, weaning conformation and weight).

Under these conditions where crude information is available on growth potential and especially since no records are available on feed efficiency and slaughter performance, the practice of performance testing appears more necessary than ever. At first, minimum numbers of recorded contemporaries are required to ensure some genetic efficiency to the programme. In France, it has been recommended to test at least 40 bulls per year (with a minimum of 15 born within 6 weeks) so that the mean genetic level of the 50p.100 upper selected bulls will be closer with a probability of 0.98, to the true genetic level of the population within which these bulls have been sampled (Table 2). The complex biological interrelationships among the classical growth criteria (growth rate, food conversion ratio, dressing percentage and carcass grade) present many implications on performance testing procedures (feeding system, age of animals, length of test) and their genetic efficiency. This has been clearly shown by Andersen (1978), Bonaiti and Béranger (1980) and Andersen et al. (1980) in a very comprehensive report of an EEC-EAAP working group. Therefore, it is not the purpose here to go back to all these considerations. Let us just limit to the main outlines of the new version of the test carried out now in France which could provide a basis of discussion. The purposes of this test (Table 3) are the following:

(i) - <u>maximum possible elimination of influences prior to the test</u>:

This has been achieved through a longer adaptation period (8 weeks) than previously practiced and following an <u>ad lib</u> feeding period of 4 weeks to equalise individual body characteristics. However, will this procedure be sufficient to carry over the main influences of the rearing conditions on account of the still late starting age (about 9 months)? Some studies from New Zealand (Dalton and Morris, 1978), based on correlation values between central performance test and progeny traits are

Table 2: Minimum numbers (M) of individuals to be performance -tested so that the genetic level of the p (p=N/M) fraction selected will be positive at a given probability α*.

Probability level (α)	Selection rate (p)								
	0.1	0.2	0.3	0.4	0.5	0.6	0.7	0.8	0.9
0.900	10	10	10	10	10	14	19	34	95
0.950	30	20	17	20	22	29	40	70	198
0.975	40	25	27	28	32	40	56	99	282
0.990	50	35	37	38	44	55	79	139	396

* the coefficient of heritability of the trait considered is assumed to be equal to 0.40.

very critical in this respect. They concluded that central performance tests are of little value in predicting the genetic value of bulls for growth traits unless bulls are collected earlier than 3-4 months of age and preferably at birth. Data from Great Britain used by Smith et al. (1979) to estimate the regression of progeny test on station performance test records were unhappily too limited (13 bulls only) to clarify the question. However, the authors in analysing individual farm records also have shown that within herd performance recording is of limited efficiency contrary to the opinions of scientists from New-Zealand. In France, estimations of the relationship "performance-progeny test" (regression of the progeny index on the individual performance) made in the past when progeny-testing concerned crossbred veal calves (75 day-weight) were equal to 0.31 and 0.34 for gain in test and 400 day-weight respectively and found to be in good agreement with expected values.

(ii) Selection for growth efficiency:
Animals are fed for 14 weeks on a restricted pelleted diet (0.72 UFV per kg) the amount of which is determined on the basis of metabolic weight and a given objective

of growth rate. This rate of growth has been chosen high enough (1500, 1400 and 1300 g per day in Charolais, Blond d'Aquitaine and Limousin breeds respectively) to allow a good expression of the genetic variability of traits related to the efficiency of growth i.e. lean tissue growth capacity and maintenance requirements in a broad sense (true maintenance requirements, heat losses related to growth and digestion efficiency).

Selecting on liveweight gain in this system should lead to an improvement of feed efficiency and carcass lean content. Trials are now in progress to check this relationship with body composition through D_2O dilution space techniques. However, selection on appetite (feed intake capacity related to the ability of consuming a large amount of roughage) is not possible in this system. Whether or not this objective should be taken into account for beef crossing on dairy populations may be questionable? In fattening systems on pastures, for instance, or if crossbred females were kept for a first calving at 2 years, it might be a problem for some breeds such as the Limousin or perhaps the Blonde d'Aquitaine. But it could be argued that this trait is also selected indirectly through growth rate of heifer progeny on pasture (progeny test of daughter groups) in beef breeds and through milk production of heifer progeny in dairy breeds.

The same question could be raised about the possible unfavourable effect of too strong a selection for lean tissue growth in the same production systems. In that case, it can be expected that crossing with dairy breeds showing a generally early maturing pattern will correct, at least partly this effect of selection. In addition, to its favourable influence on carcass traits (dressing percentage, bone to muscle ratio), live muscle conformation scored by trained experts in the standardised environmental conditions of a station, should remain an important selection criterion in order to minimise, in crossing the deficiencies of animals presenting a more and more

Table 3: Performance testing procedure applied in France for terminal beef breeds

Age (months)	Length of the period (weeks)	Stages	Feeding	Objectives
9 11	8	Adaptation	hay + concentrate restricted	Reduction of the pre-test environmental effects
12	4	Ad libitum	Concentrate ad lib + 2 kg straw	same + culling on low feed intake
15.5	14	Restricted (growth rate fixed)	Adjusted feeding on metabolic and growth requirements	selection on growth efficiency. (feed efficiency, lean growth capacity)

pronounced dairy type, detrimental to the commercial value of 8 day old or veal calves (Colleau, 1981). However, up to now it seems very difficult to determine on an objective basis the relative weights which should be given to muscling and growth traits on account of the variability existing for these traits between the different beef breeds available and the production and commercial conditions prevailing in the beef industry.

2. Selection for Easy Births

This factor appears as one of the most limiting to the practice of terminal crossing in dairy breeds due to its direct and indirect effects (labour costs for supervision and assistance, stillbirth, retained placenta, reduced subsequent fertility and milk production) on the herd productivity (Philipsson, 1977; Smidt and Huth, 1979). It is especially more urgent to pay attention to this trait as the strong selection applied after performance and progeny-test on growth potential (weaning and yearling weight) induces a significant increase in birth weight (Andersen et al., 1974; Koch et al., 1974; Foulley, 1976) and consequently in difficult births.

(i) Sire evaluation

To this respect, we have stressed (Foulley and Menissier, 1979; Philipsson et al., 1979) the interest of progeny-testing bulls for birth conditions on birth weight rather than on score or frequency of difficult births of progeny calves. The heritability of birth weight is considerably higher than that of score or frequency of dystocia (h^2=0.20 vs 0.05 in field conditions). Its genetic variation is not related to age of dam as happened for frequency of dystocia. Moreover, the correlations between, on the one hand, sire proofs for dystocia in heifers and, on the other hand, sire proofs for birth weight in heifers and cows, respectively are high and practically the same (Burfening, 1979; Philipsson et al., 1979). Then, if accurate recording of birth weight is possible, progeny-testing for easy births can be carried out not only on heifers but also on cows. It is the procedure, we have recommended in France. In addition, the set up in progeny test of given reference sires in each breed can help to control the genetic evolution of this trait over space and time. Standards have been proposed for progeny group size of young bulls and reference sires in order to achieve the optimum distribution of total number of progeny for a given accuracy (Foulley et al., 1981; Table 4). Thus, sire evaluation for birth difficulties using cow data requires, much more offspring (3 times more in conditions of Table 4) if made on the frequency of dystocia rather than on birth weight. Nevertheless, the choice of any of these methods will also depend on other practical considerations such as recording systems used according to the country and type or breed of cattle. Moreover, simultaneously taking into account birth weight and other calf traits (e.g. gestation length, morphology) seems of little practical value in improving the prediction of direct genetic effects of dystocia (Menissier et al., 1974; Burfening, 1978). However, if recorded, data on thes e traits could be useful to control their genetic evolution under the effects of selection or

even prevent any unfavourable change in them by using indices with restriction.

(ii) Selection strategy for calving ease

In fact, the choice of the beef sire breed and its selection will depend on:

. the rate of acceptable difficult calvings in milking herds;
. the maternal calving ability of the dairy breed used;
. the choice of females (especially according to age) submitted to beef crossing.

Despite existing differences among countries, breeds and production systems, the rate of difficult calvings which is economically tolerable in dairy herds can just be low (about 5 to 7 p.100 maximum for the whole herd) even though breeders generally overestimated this risk (Meijering, 1980). As far as differences between breeds are concerned, progeny test data in France show the poorer maternal calving ability of dual purpose breeds of large size and good muscling (Pie Rouge de l'Est; Brune des Alpes) in comparison with more specialsied dairy herds such as Friesian and Montbeliarde (Table 5). Within the Black and White breed, a Dutch study based on a crossbreeding experiment (Oldenbroek, 1980) reports a favourable estimate of the maternal effect on calving ability for Holstein over Dutch Friesian; but, considering the maternal component (true maternal plus half direct effect), there is only a small difference (in favour of the Dutch Friesian) between the two, due probably to a balance between simultaneous influences of Holstein genes on birth weight and pelvic opening. However, these dam breeds have not yet been compared in crossing with large size beef breeds.

As far as the choice of females submitted to terminal crossing is concerned, different procedures can be imagined, especially on account of the incidence of calving problems to be expected according to age (Philipsson, 1977; Figure 1). One possibility to take

Table 5: Calving performance of dairy and dual purpose breeds used in crossing with Charolais AI bulls (progeny test data : Foulley et al., 1975).

Dam breed	No of calves	Birth weight (kg)	Calving score (a)	Difficult births (p.100) (b)
French Friesian	3,627	42.82	1.48	5.2
Montbeliarde (c)	2,060	+1.87	-0.20	-1.4
Pie Rouge de l'Est (c)	1,934	+3.26	+0.44	+1.1
Normande (c)	376	+0.43	-0.14	-0.2
Brune des Alpes (c)	335	+2.49	-0.01	+2.4

(a) 1 : no assistance 2 : slight assistance 3 : strong assistance
4 : caesarean 5 : embryotomy.

(b) frequency of scores 3,4 and 5.

(c) In deviation from French Friesian mean.

maximum possible advantage of potentialities of large beef breeds and bulls while trying at the same time to minimise their unfavourable effect on dystocia is just to cross them to the oldest cows. Consequently, this implies keeping all female progeny from heifer matings for replacement. The natural alternative in a dairy system is to cross breed females discarded on account of their expected breeding values for milk yield, in order to reach the maximum genetic improvement in that trait along the path "dam to daughter". The first solution leads to immediate and long-term losses in milk returns but to improvement in meat. The economic comparison of these alternatives depends on the cow replacement rate, annual genetic improvement desired in milk and ratio of beef value added per crossbred over pure calf to the margin per kg of milk (Elsen and Mocquot, 1976).

Therefore, according as only mature or young females are crossbred with beef bulls, different strategies of use and selection of the latter may be applied. A very efficient way in the short term to solve these calving problems lies in a differential use of bulls for young vs mature cows on the basis of their sire evaluation for birth conditions without any selection realised on that trait (Philipsson et al., 1979). However, this procedure does not provide any long term solution to the problem, particularly in the case of continental beef breeds strongly selected on growth potential. Consequently, in these breeds combined selection for pre and postnatal growth would be absolutely necessary to protect their advantage for crossbreeding in the future. It is theoretically possible by the use of restricted selection criterion leading to the maximum genetic progress in yearling (or weaning) weight with a given expected change in birth weight compatible with the characteristics of breeds crossed and the risk of dystocia accepted (Foulley, 1976; Table 6). For instance, with beef bulls mated to mature Friesian cows, the threshold of birth weight corresponding to a risk of dystocia of 5p.100 is

Table 6: Restricted selection indices on birth weight to be applied to bulls of terminal crossing (from Menissier, Foulley and Pattie, 1979

		Relative restriction in birth weight (p.100)				
		100	75	50	25	0
Performance test						
(a) ΔG_{BW}	(kg)	0	0.064	0.127	0.191	0.254
(a) ΔG_{W400}	(kg)	3.73	4.60	5.30	5.83	6.05
	(p.100)	61.6	76.0	87.6	96.3	100.0
(b) K		-5.40 (c)	-3.68	-2.25	-0.84	+0.83
Progeny test (n=20)		0	0.080	0.162	0.241	0.322
(a) ΔG_{BW}	(kg	6.02	6.76	7.35	7.74	7.90
(a) ΔG_{W400}	(p.100)	76.2	85.6	93.1	98.0	100.0
(b) K		-5.40 (c)	-3.95	-2.53	-1.10	+0.60

(a) ΔG_{BW}, ΔG_{W400} : expected genetic change per generation in birth weight and 400 day weight respectively by selecting males with a selection intensity of 1.

Parameters used are:

h^2 = 0.10 and 0.40 as heritability of BW and W400

r_g = 0.45 and r_p = 0.09 as genetic and phenotypic correlations

σ = 5 and 30 kg for phenotypic standard deviation of BW and W400

(b) K : coefficient to apply to birth weight in the index I = W400 + K.BW

(c) : equal in the two situations to $-r_g\ h_{W400} \times \sigma_{W400} / h_{BW} \times \sigma_{BW}$

about 45 kg (Figure 1). On account of the present genetic level of the Charolais breed (44 kg in these conditions according to Menissier et al., 1981), birth weight should practically not be increased genetically anymore in this breed. In fact, no real selection on the basis of either a restricted index of culling level on birth weight has taken place in this breed. Sire evaluations for birth difficulties and/or weight are merely used to provide recommendations for differential matings of bulls. Another attempt to solve this problem without penalising too much the postnatal growth potential is to create a synthetic line combining complementary abilities of different breeds. This is what has been done by an AI unit in the southwest of France in selecting a gene pool named "Coopelso 93" constituted of Charolais (growth potential), Blonde d'Aquitaine (long bodied with an adapted morphology) and Limousin breeds (low birth weight and carcass quality).

In Friesian heifers crossbred with beef bulls, the maximum birth weight of progeny compatible with a 10p.100 probability of a difficult calving is around 38 kg which corresponds quite well with the present genetic level of the breed (Foulley, 1978; Menissier et al., 1981). Now, the Limousin breed appears as one of the few continental beef breeds able to improve in intensive production systems, significantly more beef characteristics of calves than do some British breeds such as the Angus, without increasing too much the calving problems of young females (Pourtier, 1980). Therefore, a breeding programme of a terminal Limousin strain selected for postnatal growth potential with a restriction on birth weight has been set up by a French AI unit (Table 7).

3. Selection Strategy for Terminal Sire lines used in Dairy Populations

This example leads us naturally to raise the question of what kind of terminal sire lines should be selected for dairy populations.

FIGURE 1 Relationship between frequency of calving difficulties and birth weight of calves out of Friesian dams mated to beef bulls (progeny test data)

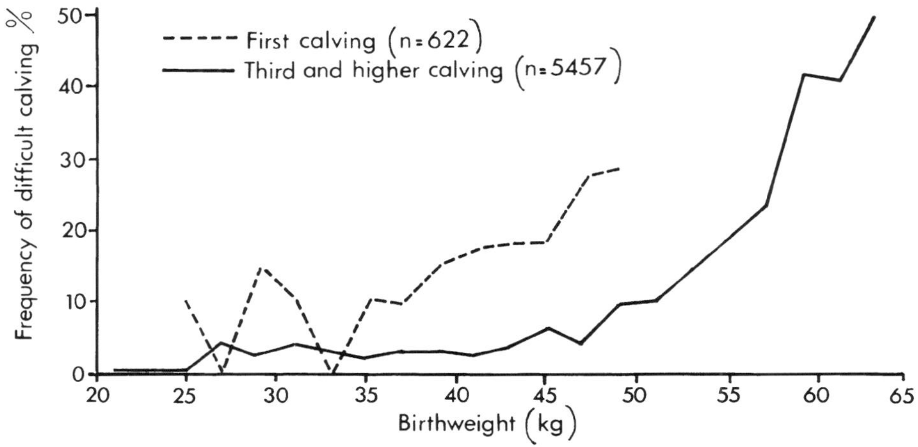

Table 7: Some results about the Limousin strain (ALPHA 16) selected with restriction on birth weight

	No of bulls	Birth weight (kg) (1)	Final weight
a. Foundation generation			
.AI bulls available in 74	95	38.6	103.8
.Bulls initially selected	15	-1.8	+ 0.9
b. Progeny test of the first 2 batches			
.Deviation selected/recorded	7/19	-0.12	+ 3.1
.Deviation selected/reference sires	7/3	-0.67	+ 4.7

(1) measured on progeny

(2) relative breeding value for 75 day weight in (a) and kg of expected progeny difference for 120 day weight (phenotypic standard deviation = 17 kg) in (b).

(i) With regard to selection goals it is obvious that they can be very numerous on account of:

. variation in, on one hand, calving performance of females crossed according to the dairy breeding type concerned and age and, on the other hand, in rate of beef crossing accepted in relation to the level of dystocia and side effects tolerable;

. multiplicity of production and commercial systems in which crossbred progeny are raised and valorised; this concerns as well proportions of 8 day old, veal and weaned calves as the ways older bulls and chiefly crossbred heifers are managed.

The first point has been previously discussed and it has been stressed by the two existing tendencies towards selection of "maximal" (for instance in Charolais) versus "minimal" strains as ALPHA 16 Limousin.

The second is just as important. It justifies to a great extent the use of all existing beef breeds in terminal crossing particularly in France considering their specific traditional areas and the corresponding diversity of types of animals marketed. In this respect, the special demand existing in France for highly muscled animals, particularly veal calves, has led to the creation by INRA of a specialised double muscled sire line essentially out of Blonde d'Aquitaine and Charolais breeds. By the way, the use of these double muscled bulls in crossing with dairy cows could solve to some extent the problem of the female fattening system (Menissier, 1980). Due to their delay in sexual maturity, heifers could reach better commercial weights in an intensive fattening system without too much fat contrary to what happens with those sired by normal bulls.

(ii) Nevertheless, should we select as many sire lines for crossing in dairy populations as existing objectives? Theoretically, on a genetical basis, it is worthwhile to select specialised lines on a limited number of traits to benefit as far as possible from heterosis and complementarities between genetic stocks of parental lines. But one critical factor to achieve an efficient selection will be the size of the selection nucleus, particularly that of females in AI breeding. In some situations (building a synthetic breed) or for some selection objectives whose side effects could not be tolerated by individual breeders (as for instance double-muscling), creating at the selection unit level, its own herd of female nucleus will appear as interesting or even absolutely necessary. As a matter of fact, this allows one to set up a better recording system, to control more efficiently selection decisions and planning of matings than in field conditions where cows are spread over many herds and breeders.

But it also costs quite a lot more for the selection unit and the size of the nucleus created will prove to be generally too limited, even though highly selected at the start, to achieve much more genetic improvement through the dam path than might be expected in a well organised recorded field population. Opening selection nucleus (James, 1977), as well as practising egg-transfer could certainly overcome to some extent this disadvantage. However, even in that case, long term genetic improvement could be jeopardised due to too small effective population size. In France, some of these factors certainly acted in difficulties encountered by those developing specialised terminal strains as might be assumed on the basis of their present value compared to that of conventional breeds (Menissier et al., 1981).

(iii) An important factor also to be considered is the importance of the demand (AI numbers) for a given objective. If it appears to be too small or unreliable in the future, it will be very difficult for an AI unit to keep on longer a specific selection for that line.

This is what is now happening to some extent in France for the five AI selection units involved in selecting terminal beef bulls within the three main beef breeds, as in specialised strains (17662 first AI in 1979).

As terminal crossing has regularly decreased over the last 10 years partly because of disappearance of many small farms raising hardy and traditional dual purpose cattle, partly owing to the development of Black and White cattle in more and more specialised dairy units, the selection units are now inclined to put together in the same selection programme buils for terminal crossing and bulls intended to produce purebred suckling cows. These terminal bulls will be chosen after a performance and a very simple field progeny-test on opportunity objectives but no real specific selection will be made for them in the long term except that relevant to the selection objectives for suckling herds.

Conclusion

Under these conditions, it becomes urgent to develop means for protecting the future of terminal sire breeds which have been selected up to now (for about 20 years) in very efficient integrated programmes.

First of all, prospects of beef crossing should be rationally analysed for the mid and long term, more especially for the time when the dairy specialisation will be mostly accomplished (population size at equilibrium; proportion of Holstein genes fixed) and then taking into account the corresponding consequences of that phenomenon on meat production.

Contrary to most present opinions now propagated in France, we are convinced that some place will remain for terminal sire lines specifically selected for crossing on dairy populations. How many? Certainly not as many as the breeds selected and strains developed since the 70's, but at least 2 relatively different types. One might be selected mainly for muscling but also for a good growth potential and with a slight restriction on birth weight. This strain could be used on mature cows in herds of average milk level accepting some amount of dystocia, but valorising quite well crossbred progeny in meat production of high grade. The other one, specifically selected for a very low rate of calving problems, but with enough growth and particularly muscling, would be intended to breed young females or, more generally those of herds where unfavourable effects of beef crossing must be minimised.

This task has certainly to be undertaken at the national or even perhaps, European level, for all countries trying to obtain a better balance between meat and milk production: .

References

Andersen, B.B., 1978 Animal size and efficiency, with special reference to growth and feed conversion in cattle.

Anim. Prod., 27, 381-391.

Andersen, B.B., Fredeen, H.T., Weiss, G.M., 1974. Correlated response in birth weight, growth rate and carcass merit under single-trait selection for yearling weight in Beef Shorthorn Cattle.

Can. J. Anim., 54, 117-125.

Andersen, B.B., De Baerdemaeker, A., Bittante, G., Bonaiti, B., Colleau, J.J., Fimland, E., Jansen, J., Lewis, W.H.E., Politiek, R.D., Seeland, G., Teehan, R.J., Werkmeister, F., 1980. Performance testing of bulls in AI.

Commission on cattle production, 31 EAAP meeting, Munich, FRG, 1-4 Sept. 1980.

Bonaiti, B., Béranger, C., 1980. Some considerations about performance testing of beef cattle in France.

Commission on cattle production. 31 EAAP meeting, Munich, FRG, 1-4 Sept. 1980.

Burgening, P.J., Kress, D.D., Friedrich, R.L., Vaniman, D.D., 1978. Phenotypic and genetic relationships between calving ease, gestation length, birth weight and pre-weaning growth.

J. Anim. Sci., 47, 595-600.

Burfening, P.J., Kress, D.D., Friedrich, R.L., Vaniman, D.,1979. Ranking sires for calving ease.

J. Anim. Sci., 48, 293-297.

Cartwright, T.C., 1970. Selection criteria for beef cattle for the future.

J. Anim. Sci., 30, 706-711.

Colleau, J.J., 1981. Practical aspects and techniques in taking account of the slaughter value of their progeny in indexing bulls for milk production.

EEC seminar on "Beef production from the dairy herd", Dublin, Ireland, 13-15 April, 1981.

Cunningham, E.P., 1974a. Breeding goals for beef cattle.

Ann. Genet. Sel. anim., 6, 219-226.

Cunningham, E.P. 1974b. The economic consequences of beef crossing in dual-purpose or dairy cattle populations.
Livest. Prod. Sci., 1, 133-139.

Cunningham, E.P., McClintock, A.E., 1974. Selection in dual purpose cattle populations : effect of beef crossing and cow replacement rates.
Ann. Genet. Sel. anim., 6, 227-239.

Dalton, D.C., Morris, C.A., 1978. A review of central performance testing of beef bulls and of recent research in New Zealand.
Livest. Prod. Sci., 5, 147-157.

Elsen, J.M., Mocquot, J.C., 1976. Optimisation du renouvellement des femelles dans les troupeaux laitiers soumis au croisement terminal.
Ann. Genet. Sel. anim., 8, 343-356.

Foulley, J.L., 1976. Some considerations on selection criteria and optimisation for terminal sire breeds.
Ann. Genet. Sel. anim., 8, 89-101.

Foulley, J.L., 1978. L'amelioration genetique de la race Limousine : bilan de la selection des taureaux utilises en insemination artificielle.
Elevage et Insemination, No. 167, 9-23.

Foulley, J.L., Menissier, F., Gaillard, J., Nebrada, A.M., 1975. Aptitudes maternelles des races laitieres, mixtes, rustiques et a viande pour la production de veaux de boucherie par croisement industriel.
Livest. Prod. Sci., 2, 39-49.

Foulley, J.L., Menissier, F., 1979. Selection for calving ability in French beef breeds. In Hoffman, B., Mason, I.L., Schmidt, J., calving performance and early viability of the calf.
Cur. Top. Vet Med. Anim., Sci., 4, 159-176.

Foulley, J.L., Schaeffer, L.R., Song, H., Wilton, J.W., 1981. Progeny group size in an organised progeny test programme of AI beef bulls using reference sires in France.
Can. J. Anim. Sci., (submitted).

Frebling, J., Gaillrd, J., Vissac, B., 1972. Efficacite réelle et optimum du choix des taureaux de race a viande pour le croisement industriel.
In Bull. tech. Dep. Genet. anim., (Inst. nat. Rech. agron. Fr.), 15, 23-55.

Gaillard, J., Foulley, J.L., Menissier, F., 1974. Observations sur l'efficacite du choix sur ascendance paternelle et performances individuelles des taureaux de races à viande destines au croisement terminal.
1er Congr. Mond. Genet. appl. Elev. Madrid, 3, 877-887.

Hill, W.G., 1971. Investment appraisal for national breeding programmes.

Anim. Prod., 13, 37-50.

James, J.W., 1977. Open nucleus breeding systems.

Anim. Prod., 24, 287-305.

Koch, R.M., Gregory, K.E., Cundiff, L.V., 1974. Selection in beef cattle. II Selection response.

J. Anim. Sci., 39, 459-470.

Lindhe, B., 1976. Beef breeds for commercial crossing.

EEC seminar on "Optimisation of cattle breeding schemes", Dublin, Ireland, 26-28 Nov. 1975, 107-119.

Meijering, A., 1980. Beef crossing with Dutch Friesian cows : model calculations on expected levels of calving difficulties and their consequences for profitability.

Livest. Prod. Sci., 7, 419-436.

Menissier, F., 1980. Advantage of using double-muscled sires in crossbreeding and selection of specialised double-muscled sire line in France.

EEC seminar on "Muscle hypertrophy in cattle", Toulouse, France, 10-12 June, 1980.

Menissier, F., Bibe, B., Perreau, B., 1974. Possibilities d'amelioration des conditions de velage par selection. II L'aptitude au velage de trois races à viande francaises.

Ann. Genet. Sel. anim., 6, 69-90.

Menissier, F. Foulley, J.L., Pattie, W.A., 1979. The calving ability of the Charolais breed in France and its possibilities for genetic improvement.

Meet. Techn. Com. Int. Fed. Ass. Char. Breed., Vichy, France, 6-7 Sept. 79, 39pp.

Menissier, F. Sapa, J., Foulley, J.L., Frebling, J., Bonaiti, B., 1981. Comparison of different sire breeds crossed to Friesian cows : preliminary results.

EEC seminar on "Beef production from the dairy herd", Dublin, Ireland, 13-15 April 1981.

Mocquot, J.C., 1972. Recherches sur la rentabilite optimale du schema de selection des males de races a viande pour une production de veaux et (ou) de jeunes bovins.

EAAP meeting, Verone, Italie, Oct. 1972.

Mocquot, J.C., Foulley, J.L., 1973. Recherche des conditions de rentabilite d'un schema de selection d'une souche de bovins destines au croisement de premiere generation pour la production de veaux de boucherie.

Ann. Genet. Sel. anim., 5, 189-209.

Oldenbroek, J.K., 1980. Breed and crossbreeding effects in a crossing experiment between Dutch Friesian and Holstein Friesian cattle.

Livest. Prod. Sci., 7, 235-241.

Philipsson, J., 1977. Studies on calving difficulty, stillbirth and associated factors in Swedish cattle breeds. VI Effects of crossbreeding.

Acta Agr. Scand., 27, 58-64.

Philipsson, J., Foulley, J.L., Lederer, J., Liboriussen, T., Osinga, A., 1979. Sire evaluation standards and breeding strategies for limiting dystocia and stillbirth.

Livest. Prod. Sci., 6, 111-127.

Pourtier, D., 1980. Comparaison de trois races paternelles de croisement sur des genisses de races a viande pour la production de jeunes bovins de boucherie.

I.N.R.A., Jouy-en-Josas, 50pp (ronéoté).

Smidt, D., Huth, F.W., 1979. Survey of the incidence of calving problems, calf mortality, and their economic importance : dairy and dual-purpose cattle. In Hoffmann, B., Mason, I.L., Schmidt J., Calving performance and early viability of the calf.

Cur. Top. Vet. Med. Anim. Sci., 4, 3-29.

Smith, C., Steane, D.E., Jordan, C., 1979. Progeny test results on Hereford bulls weight-recorded on the farm.

Anim. Prod., 28, 49-53.

DISCUSSION ON DR. J.L. FOULLEY'S PAPER

Chairman: Dr. R. Thiessen

Cunningham: Are there any advantages in reducing gestation length as an alternative to reducing birth weight ? The potential advantage I see is that gestation length can be measured more cheaply and ergo greater numbers.

Foulley: I have looked at this, and it is relatively inefficient to use gestation length to predict calving difficulties. There appears to be no advantage over using birth weight alone for this purpose. Greater numbers of progeny are needed, and there will be a larger decrease in gestation length itself, which may not be desirable.

Bech Andersen: We have also done some work on the problems concerning gestation length and we have found a high heritability for this trait, but also a very low within breed genetic variation. It is possible to reduce birth weight by indirect selection to reduce gestation length, but to obtain a reduction of 1 kg in birth weight, gestation length needs to be reduced by 3 - 4 days. Direct selection on birth weights to obtain a 1 kg reduction would result in a decrease in gestation length of only 1 - 2 days. We have not recommended that our breeders include gestation length in their selection programmes for this reason. One problem might be that selection for decreased gestation length could lead to an increased frequency of immature calves for example.

Jansen: What is the effect of restricting concentrate during the first 8 weeks on the subsequent 4 weeks of *ad libitum* feeding in the performance test regime you outline ?

Foulley: It is not a constant restriction, as there is a progressive increase in the level of feeding over the 8-week period.

Langholz: Do you have any information about the relationship between gestation length and incidence of caesarians ? We found in our experiments that calves delivered by caesarian had gestation lengths that were 34.8 days longer than average. It is these extreme values of gestation length that we should pay pay attention to, both when analysing data and when making breeding decisions.

Foulley: I am not familiar with this occurrence of such extreme gestation length values.

Jansen: What is the effect of using only linear regressions in these calculations, for example, the relationship between dystocia and gestation length ?

Foulley: It depends on the range within which you are working. The relationship between dystocia and gestation length is not linear, but if you are working within a particular area of the relationship, it makes very little difference.

Oostendorp: You were mentioning the development of a Limousin strain that could be used on heifers without causing calving difficulties. I would like to know what the experience is up to now with regard to calving difficulties and beef production traits ? Further to this, is there a possibility that Professor Romita could use this strain of Limousin in his experiments with heifers that calve once.

Foulley: I refer you to Table 7 for the results using the Alpha 16 Limousin strain. Bulls from this strain are commercially available, so that Professor Romita can avail of them if he so desires.

CHAIRMAN'S COMMENTS AT CLOSE OF SESSION II

Chairman : Dr. R.B. Thiessen

The papers in this session on "Beef breeds for crossing in dairy herds" detailed results in two important areas. The first was the effect on calving difficulty and calf mortality when the large continental sire breeds are crossed to dairy breeds. A report of interest was that of pelvic shape and size in relation to body weight in the Jersey breed, and the relative ease of calving when crossed to Charolais sires.

The second area of consideration was the growth, food intake, carcass composition and feed efficiency of crossbred slaughter calves. There was general agreement of results between countries when the same end points at slaughter were considered. Conflicting results between countries may well be explained by adjustment to a common end point. Because of the differing emphasis in various countries on carcass weight, degree of finish and conformation, biologically interpretable end points such as degree of maturity or associated finish (fatness) would be desirable. Serial slaughter techniques would allow adjustment of end points by regression analysis. Economic considerations could then be imposed at different commercial end points. Useful strategies for selection of sire breeds were also considered.

INFLUENCE OF ECONOMIC WEIGHTS AND POPULATION STRUCTURE ON SELECTION RESPONSE OF MILK AND BEEF TRAITS

D. Fewson

Universitat Hohenheim,
Institut für Tierhaltung und Tierzüchtung,
GarbenstraBe 17, D 7000 Stuttgart 70.

Summary

The effect of economic weights and population structure on selection response of milk and beef traits was investigated by aid of model calculation.

The economic weights of dairy traits calculated on national level are lower than on farm level. The reverse is true for beef traits. If the breeding objective is determined on farm level one gets a satisfactory selection response on both farm level and on national level. Therefore, the breeding objective should be determined on farm level. Otherwise the breeder runs the risk that the farmer buys his breeding stock somewhere else.

The optimal proportion of active breeding population proves to be around 50 to 60%. In this region there are combined a high profit with a satisfactory genetic gain.

Compared with milk recording the testing system for beef traits is underdeveloped. The extension of performance testing stations for bulls as well as the introduction of progeny test of bulls and performance test of cows in the field, yield a higher genetic gain and a higher profit. Moreover, these breeding activities seem to be an effective tool to balance the genetic gain of milk and beef traits.

Introduction

The balance of genetic gain between milk and beef traits is mainly influenced by the economic weights of these traits and the population structure. The economic weights bring about changes in both the importance of the single traits in the aggregate genotype and in the estimated general breeding value. In this way the economic weights effect the genetic gain of different traits.

All factors of population structure influence more or less the composition of genetic gain. In this respect there will be discussed the proportion of active population or milk recording, the capacity of performance testing stations for young bulls and the different testing methods for beef production.

The results which are presented here, originate from a study of Henze et al. (1980) regarding the efficiency of testing methods in animal breeding. For modelling studies we used an extended version of the programme of Niebel (1974).

Economic weights

Parameters on farm or national basis

The economic weight of a trait depends on the marginal change in profit and the number of discounted expressions. The second term will not be discussed here. The marginal change in profit has to be estimated by different methods.

- If the limited resources of the farms and the prices and costs are not influenced by an improvement of a trait, the marginal change in profit may be calculated by the difference between marginal income and marginal costs. This may be true for management traits like milk flow or quality traits like lean meat content.
- If the limited resources of the farms - forage, labor, capital - are influenced, one has to use a model for the single farms.
- If - however - an improvement of a trait influences the limited resources of the region considered it's suitable to use a sectoral model. Limited resources of a region may be related to sale capacity - overproduction - or to production capacity. Such sectoral models are necessary for most traits in cattle breeding.

For our purpose we calculate the marginal change in profits of the single traits on farm level and on national level as well. For the computations on farm level we use the profit of the single farms as criterion of maximisation. There are no limitations on excessive supply. On the other hand we take the national net income as basis for calculations. Furthermore, we introduce certain limits of available supply. An excessive supply has to be sold on world markets at lower prices. The calculated economic weights under both conditions for one genetic standard deviation are listed in Table 1.

Table 1 Economic weights for one genetic standard deviation (DM)

Trait	Farm level	National level
Milk yield	23,34	13,62
Fat yield	24,30	15,19
Milk flow	7,78	7,78
Daily gain	32,04	34,26
Food conversion	15,90	17,35
Growth capacity	5,08	25,38
Lean meat content	16,85	16,85
Calving interval	3,74	2,54

As expected the economic weights of dairy traits are highly reduced on national level, this is related to the excessive supply of milk. On the other side the importance of beef traits on national level is increased. This is especially true for growth capacity. The reason for this is the demand for the available supply for beef.

Composition of genetic gain

For the model calculations on the influence of the economic weights of beef and dairy traits we assume a waiting-bull system with beef performance testing of young bulls on station and progeny testing for milk yield. The population structure may be characterised by the following factors:

- Population size 1 000 000 cows
- Proportion of active breeding population 0.30
- Proportion of cows inseminated with young bulls 0.30
- Capacity of performance testing station 1 000 places
- Proportion of selected bulls under performance test on beef traits 0.40
- Number of milk recorded daughters for young bulls 63
- Number of bull sires per year 10

In Table 2 is shown the influence of economic weights on farm or national level on the composition of genetic gain.

Table 2 Composition of genetic gain depending on the calculation of economic weights on farm or national level

Trait		Farm level	National level
Milk yield	kg	+ 42,7	+ 33,5
Fat yield	kg	+ 1,80	+ 1,42
Milk flow	kg/min	+ 0,011	+ 0,008
Daily gain	g	+ 6,70	+ 7,08
Food conversion	kStE/kg	- 0,015	- 0,012
Growth capacity	kg	+ 3,68	+ 4,21
Lean meat content	%	- 0,034	- 0,021
Calving interval	Tg	+ 0,43	+ 0,31
Total genetic gain	DM	+ 12,65	+ 12,49

If we use a breeding objective on national level the genetic gains of dairy traits are reduced by more than 20%. In relation to this the decrease of lean meat content and the increase of calving interval is diminished. This depends on the assumed negative genetic correlation between dairy traits and lean meat content or calving interval. As expected, the genetic gain of beef traits is improved by using economic weights on a national basis.

The monetary genetic gain of milk and beef traits compensate each other. Therefore, the total genetic gain is nearly the same for economic weights on a farm and national level.

Determination of breeding objective and income

Up to now we used the same economic coefficients of the traits for determination of the breeding objective and the income. But there are two questions still to be answered. What happens to the national income if the breeding objective is determined on farm level? And the reverse question: What happens to the income of the farmer if the breeding objective is determined on national basis? Table 3 gives a comparison of the four methods of calculation.

Table 3 Determination of breeding objective and income on farm or national level

Breeding objective		Farm level		National level	
Income determination		Farm level	National level	National level	Farm level
Genetic gain	DM/Cow, Year	12,65	12,35	12,49	10,03
Milk traits	%	65,7	42,2	32,9	52,7
Fattening ability	%	45,7	69,0	74,4	55,6
Lean meat content	%	- 9,9	-10,2	- 6,4	- 7,2
Calving interval	%	- 1,5	- 1,1	- 0,8	- 1,2
Net income	DM/Cow	71,96	71,90	73,90	53,80

If the breeding objective is determined on farm level one gets the absolute genetic gains of single traits as shown in the second column of Table 2. Now, if the income is calculated alternatively on farm or national level milk and beef traits compensate each other; the net income is nearly the same.

We get a very similar solution for a situation, in which both the breeding objective and the income are determined on national level. The proportion of milk traits of the genetic gain is somewhat lower, the porportion of beef traits somewhat higher. Therefore, we can conclude that - from national point of view - the reduction of income is unimportant, if the breeding objective is determined on farm level.

Now we have to look at the income of the farmer if the breeding objective is determined on national level. In this situation one has to weight the absolute genetic gains of single traits as shown in the third column of Table 2 with the economic coefficients on farm level of Table 1. The result is a drastic reduction of net income.

The buyer of breeding products is a farmer, and he has to calculate with prices and costs on farm level. On the other hand the breeders are competitors, and this is true also for breeding associations. Therefore, it will also be necessary for the future to determine the breeding objective on farm level. Otherwise the breeder runs the risk, that the farmer buys his breeding stock somewhere else. In such a situation - as in E.E.C. - it seems necessary that the relations of the more or less political prices on farm level have to be changed in relation to the public interest.

Population structure

Proportion of active breeding population

In the calculations mentioned above we used a fixed proportion of active breeding population of 30%. Now we look at the influence of this factor and vary the proportion of active breeding population in the wide range from 10 the 100%. The breeding objective and the income are determined on farm level. In Table 4 the results are summed up.

The total genetic gain increases with the proportion of active breeding population. This increase is not linear but declining. As expected the composition of genetic gain changes too. The percentage of milk traits increases slightly with the proportion of active breeding population. On the other hand the percentage of fattening ability is reduced.

The development of gross income is nearly the same as that of the genetic gain. However, the increase of the breeding expense is almost linear. This depends on the costs of milk recording. The declining increase of gross income and the linear increase of expense result in a maximum for the net income. In our situation the maximum is reached if the proportion of active breeding population is around 30%. But the shape of the curve of net income is very flat. Now, if limits are tolerated for the net income within 5% of its maximum, one gets a range of acceptable proportions of active breeding population between 10 and 60%. Within this range it seems useful to prefer a population structure with a high genetic gain. Then the optimum of the proportion of active breeding population will be around 50 to 60%. This reflects the actual situation in our country.

Table 4 Influence of the proportion of active breeding population on genetic gain and income

Characteristic		Proportion of active population in %									
		10	20	30	40	50	60	70	80	90	100
Genetic gain DM/Cow, Year		11,04	12,10	12,65	13,00	13,26	13,46	13,63	13,77	13,90	14,02
Milk traits	%	61,0	64,4	65,7	66,4	67,0	67,3	67,5	67,7	68,0	68,1
Fattening ability	%	49,6	46,8	45,7	45,1	44,6	44,4	44,2	44,1	43.9	43.7
Lean meat content	%	- 9,3	- 9,7	- 9,9	-10,0	-10,1	-10,2	-10,2	-10,2	-10,3	-10,3
Calving interval	%	- 1,3	- 1,4	- 1,5	- 1,5	- 1,5	- 1,5	- 1,5	- 1,5	- 1,5	- 1,5
Gross income	DM/Cow	76,0	83,2	87,7	91,1	94,0	96,6	99,1	101,4	103,6	105,8
Expense	DM/Cow	7,6	11,7	15,7	19,8	23,9	28,0	32,1	36,2	40,3	44,4
Net income	DM/Cow	68,4	71,6	71,9	71,2	70,1	68,6	67,0	65,2	63,3	61,4

Capacity of performance testing stations

Milk recording is extended to all breeding herds and to a proportion of commercial flocks. In comparison to this the recording system of beef traits is underdeveloped. In some populations there is only a performance test of young bulls in stations with supplementary bull judging. In this section the capacity of performance testing stations is varied and the related influences are discussed.

Table 5 Influence of the capacity of performance testing station on genetic gain and on income

		Capacity of performance testing stations		
		500	1000	2000
Genetic gain	DM/Cow, Year	12,17	12,97	13,54
Milk traits	%	65,0	62,2	59,8
Fattening ability	%	46,3	48,7	50,8
Lean meat content	%	- 9,9	- 9,6	- 9,3
Calving interval	%	- 1,4	- 1,3	- 1,3
Gross income	DM/Cow	81,6	89,9	96,3
Expense	DM/Cow	12,4	12,8	13,6
Net income	DM/Cow	69,2	77,1	82,7
Proportion selected		0,40	0,20	0,10

If the requirement of young bulls for breeding is fixed, an extension of the performance testing station results in a reduced proportion selected out of all performance tested bulls. This yields a higher genetic gain. The composition of genetic gain changes in the expected direction, lower proportion of milk traits and higher proportion of beef traits. As the improvement of gross income is higher than the increase of expense, an extension of the testing station leads to an improvement of net income too. The present capacity of testing stations is below its optimum.

Testing method for beef production

The effect of different testing methods for beef traits on

genetic gain and profit are investigated. For bulls we assume either no test of beef traits, performance tests on station (E) or progeny test on specialised commercial farms with 20 sons(N). For cows we accept alternatively no test or performance test on farms (E). As heritabilities for beef traits we use figures of 0,35 and 0,15 for station and field test respectively.

Table 6 Influences of testing methods for beef production on genetic gain and income

Characteristic		Beef testing for cows				
		-	-	-	E	E
		Beef testing for bulls				
		-	E	N	E	E + N
Genetic gain DM/Cow, Year		9,34	13,4	13,27	13,75	15,20
Milk traits	%	109,6	66,6	60,5	60,5	50,2
Fattening ability	%	13,7	45,7	50,1	50,2	57,7
Lean meat content	%	- 20,7	-10,9	- 9,5	- 9,4	- 7,0
Calving interval	%	- 2,5	- 1,4	- 1,2	- 1,3	- 1,0
Gross income	DM/Cow	62,0	94,0	94,9	99,0	109,7
Expense	DM/Cow	13,0	14,4	13,6	15,3	15,8
Net income	DM/Cow	49,0	79,5	81,4	83,6	93,9

If there is no beef testing at all, genetic gain and profit are unsatisfactory. Moreover, milk traits predominate with respect to genetic gain. All assumed types of beef testing proved worthwhile. The highest genetic gain and the highest net income are realised if the different methods of beef testing are combined. Such a breeding plan guarantees a nearly balanced genetic gain of milk and beef traits.

References

Henze, A., Zeddies, J., Fewson, D., and Niebel, E., 1980: Leistungsprüfungen in der Tierzucht. Nutzen-Kosten-Untersuchung über Leistungsprüfungen in der tierischen Erzeugung, dargestellt am Beispiel der Milchleistungsprufung beim Rind.

Schriftenreihe des Bundesministeriums fur Ernährung, Landwirtschaft und Forsten. Reihe A: Landwirtschaft - Angewandte Wissenschaft, Heft 234.

Landwirtschaftsverlag Münster - Hiltrup.

Niebel, E., 1974: Methodik der Zuchtplanung fur die Reinzucht beim Rind bei Optimierung nach Zuchtfortschritt und Zuchtungsgewinn.

Dissertation Hohenheim.

DISCUSSION ON PROF. DR. D. FEWSON'S PAPER

Chairman: Prof. A. Neimann-Sorensen

Oldenbroek: Can you use the same economic weights for the different breeds in Germany ?

Fewson: No, because the means differ between populations.

Cunningham: What are the differences between the three beef traits described in Tables 2 and 3 : daily gain, growth capacity and fattening ability ?

Fewson: Daily gain is self-explanatory; growth capacity means mature weight; fattening ability is a combination of these two traits with feed conversion efficiency.

Langholz: I would like to make a comment on that point, and underline the need to differentiate between growth capacity and growth intensity. Growth capacity is identical with mature size and weight, whereas growth intensity, measured by daily gain, is the speed of obtaining the final size. The Italian breeds, for example, have a high growth capacity but are relatively slow growing, whereas Charolais have a high growth capacity and a high growth intensity.

Jansen: You put a lot of emphasis on growth capacity at a national level in order to reach higher slaughter weights. Would it not be better to modify this by (a) restricted feeding, or (b) the use of crossbreeding as an approach, rather than increasing the mature size of females.

Fewson: (a) We are assuming that the numbers of cows will be reduced drastically in the future. Therefore, it seems necessary that the fattening animals must reach higher weights. This could also be achieved by a lower feeding intensity level.

(b) Yes, it would be helpful to use bulls with high growth capacity for crossbreeding in dairy herds.

PROGENY TESTING OF DAIRY BULLS FOR MEAT PRODUCTION AND AGGREGATION WITH DAIRY INDICES

J.J. Colleau

Station de Genetique quantitative et appliquee
INRA - 78350 JOUY-en-JOSAS - France

Summary

Performance testing of bulls on meat criteria before that on milk production is the method used in France to improve or maintain the breeding value of dairy breeds for these criteria. However, in several cases, progeny testing would be useful.

A survey is made of the first two progeny testing field trials performed in France. On average, the value of the genetic parameters is favourable. The practical methods for combining milk and meat indices in the evaluation of dairy bulls are discussed.

Introduction

Performance testing of bulls on the station before their testing for milk production is the official method used in France to evaluate and select on meat production. However, in several cases progeny testing would be useful : for example, for giving information on bulls directly tested for milk production (imported bulls) or for giving more accurate details on bulls likely to be used to supply new young bulls. In addition, progeny testing would provide the breeders with results on service bulls more easy to understand or to use because they can be expressed in commercial units. For all these reasons, a series of trials was started in France some years ago in order to investigate the feasibility and accuracy of this method. The purpose of the present paper is to give a survey of the first two trials and also to express some ideas about the method of aggregating milk and meat indices.

Survey of Progeny Testing Trials

These trials concern the Black and White breed because of the problems observed in the field of meat production as a consequence of the cross breeding with Holstein-Friesian strains rapidly expanding in France (COLLEAU 1978; COLLEAU et al., 1980).

Trial 1

In this trial, the main purpose was to examine the value for genetic objectives of the data coming from commercial organisations selling young calves at the age of 8-20 days. Good results were expected from one of these units because of the large number of classes (9) used to classify the animals on meat conformation and because of the high agreement between classifiers.

Genetic investigations were therefore conducted on the following variates:

- weight of the calf at selling
- classification
- actualized selling price

The last variate is the selling price calculated as if the calves were sold at the same date by using the selling grid which is a rather complex function of weight and classification. The first results of the genetic parameters calculated within sire genetic type are encouraging since the heritability values are ranging from 0.20 to 0.40 (See Table 1). In the selling grid, a great importance is imputed to classification and consequently this criterion is highly correlated with the overall price. Some people will consider that all these values are rather high, but it must be remembered that the moment of selling is close to birth and that the heritabilities of birth parameters are known to be high.

Trial 2

In this second trial, the purposes were more ambitious:

- study of the genetic parameters around 8 days with a much narrower age range owing to special calf collections on the farms.
- study of the genetic parameters for veal production (4 months, 105 kg carcass weight) using contracts with farmers including very simple measurements and no constraining

Table 1 Genetic parameters obtained from the first progeny testing trial on meat production at 8 days (Yonne-Loiret area; 1974-1978 - 3796 calves from 70 sires).

Diagonal = $\hat{h}^2 \pm \hat{\sigma}(\hat{h}^2)$

Above diagonal = $r_G \pm \sigma(\hat{r}_G)$

Variate	(1)	(2)	(3)
Selling weight (1)	0.22 ±0.55	0.04 ±0.15	0.51 ±0.11
Selling classification (2)		0.43 ±0.08	0.87 ±0.04
Actualised selling price (3)			0.27 ±0.05

Evaluation within genetic type of sire : Friesian, 50% Holstein; 100% Holstein

management system, especially involving group slaughterings.

- evaluation of the genetic relaticnships between performance at 8 days and performances for veal production.

The results (Table 2) obtained at 8 days agree rather well with those obtained in Trial 1. The criteria measured at 8 days, especially the classification, are heritable and might allow a systematical evaluation of the bulls. The coefficients of heritability found for veal production are not as high as might be expected but it must be remembered that the system used was not very strict.

The figures concerning the genetic relationships between performances at 8 days and performances at slaughter may be considered as very high for overall selling values and very low for the components of the overall price (weight and classification). These values are somewhat contradicting each other. However, all the correlation coefficients are known with a high sampling standard deviation (around 0.25) and consequently, the study is not conclusive enough on that point. A third

FIGURE 1 Relationships between the milk production index (AVM index) and the 8 day index for bulls with at least 30 progeny at 8 days

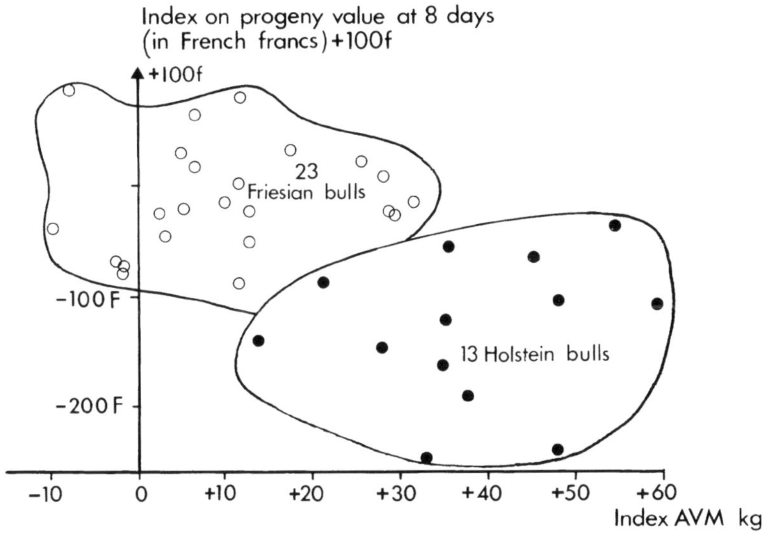

$$\text{AVM} = \text{average valuable matter} = \frac{\text{fat yield} + 1{\cdot}21 \text{ proteinyield}}{2}$$

Table 2 Genetic parameters obtained from the second progeny testing trial on meat production (at 8 days and for veal production).

Diagonal = $h^2 + (h^2)$ Finistere area : 1978-1980

Above diagonal = $r_G \pm (r_G)$

			8 DAYS			VEAL PRODUCTION				
			(1)	(2)	(3)	(4)	(5)	(6)	(7)	(8)
8 Days (a)	Selling weight	(1)	0.27 ± 0.09	0.43 ± 0.26	0.30 ± 0.24	0.09 ± 0.23	0.24 ± 0.24	-0.09 ± 0.32	0.27 ± 0.23	0.20 ± 0.30
	Selling classification	(2)		0.41 ± 0.11	0.74 ± 0.29	0.06 ± 0.21	0.07 ± 0.26	0.05 ± 0.28	0.39 ± 0.22	0.07 ± 0.29
	Actualised selling price	(3)			0.23 ± 0.09	0.60 ± 0.19	0.68 ± 0.22	0.40 ± 0.34	0.75 ± 0.18	0.79 ± 0.25
VEAL PROD-UCTION (b)	Liveweight at slaughter	(4)				0.40 ± 0.11	0.95 ± 0.03	0.21 ± 0.31	0.70 ± 0.11	0.97 ± 0.06
	Carcass weight	(5)					0.20 ± 0.09	-0.04 ± 0.37	0.78 ± 0.10	0.94 ± 0.04
	Carcass grading	(6)						0.15 ± 0.08	0.32 ± 0.33	0.29 ± 0.50
	Carcass compacity	(7)							0.36 ± 0.10	0.86 ± 0.07
	Carcass value	(8)								0.15 ± 0.08

(a) Evaluation after correction for:
- Genetic types of sire : Friesian and 25% Holstein; 50% Holstein; 75 and 100% Holstein
- n° lactation of the mother : 1,2,3 and +
- age at selling ($\bar{x}$ = 9 days)

(b) Evaluation after correction for:
- Genetic types of sire : see above
- n° lactation of the mother : 1,2,3 and +
- fattening unit (two)
- age at slaughter ($\bar{x}$ = 125 days).

(3) The price is calculated as a deviation relative to the average selling price for Friesian calves in the same area during the same month of selling.

(8) Value evaluated from the carcass weight and from a fixed commercial scale depending on carcass classification.

trial is therefore in progress to obtain further information about that topic.

Utilisation of progeny testing in practice

An efficient way to effectively use progeny testing in dairy breeds seems to be the proposal of indices combining milk and meat production criteria to dairy producers and breeders. To that end, the discounted gene flow method for evaluating the overall economic contribution of each dairy AI bull is a well known possibility (CUNNINGHAM and MC CLINTOCK, 1974).

The first question is to determine the economic function that would be evaluated or selected (breeding objective). In a first approach, only a few questions would arise among dairy breeders or dairy producers if that function would represent the overall profit made by the dairy producers by selling milk, young calves at 8 days and culled cows or heifers for meat. That function differs from that representing the overall profitability at a national level, which would integrate fattening and carcass characteristics of young cattle.

Using French demographical parameters, numerical studies over a 15 years period showed that the breeding objective is roughly proportional to the expression :

$$H = G_1M_1+0.5\ G_2M_2 + 0.25\ G_3M_3$$

where G_1, G_2, G_3 describe the genetic values for milk production, 8 days calf production and culled cow production respectively, the quantities M_1, M_2, M_3 being the corresponding margins per unit of genetic value.

Thus, the value of the breeding objective must be predicted by using the bull indices for milk production or for calf values at 8 days (real or predicted from slaughter results if the station testing is accompanied by a much higher heritability). Of course, a good knowledge of the genetic and economic parameters is required. The importance of both the genetic correlation between value at 8 days and value at culling is clearly predominant (Table 3) and consequently must be known rather well. The results from the first progeny testing trial (Figure 1) show that the value of the genetic correlation between milk production and value at 8 days is likely to be close to 0 (at least at a within population level which is the relevant one since the Friesian bulls have great chances to be eliminated in the long run). On the other hand, an intermediate value for the correlation between value at 8 days and value at culling is to be expected. Consequently, the best prediction of H, the aggregate merit of a bull, seems to be centred around the expression $\hat{H} = \hat{G}_1 M_1 + 0.60\ \hat{G}_2 M_2$ where $\hat{G}_1$, $\hat{G}_2$ are the respective indices for the milk production criterion and the commercial value of the progeny at 8 days.

An experimental application of this index was made within a population, especially within the Holstein bull population and showed that some bulls were not fit to be used as fathers to bulls although they were used as such in some areas in France. This fact emphasises the necessity of putting stress on meat production criteria in performance testing and in progeny testing as well.

General evaluation of the individual bull within and between populations was also made by using the following combined indices:

$$\begin{aligned} \hat{H}_{ij} &= H_i + (\widehat{H_{ij} - H_i}) \\ &= G_{1',i} M_1 + 0.5\ G_{2',i} M_2 + 0.25\ G_{3',i} M_3 \\ &+ (\widehat{G_{1,ij} - G_{1',i}})\ M_1 + 0.60\ (\widehat{G_{2ij} - G_{2',i}}) M_2 \end{aligned}$$

Where i = population number.

j = bull within population number.

$G_{1',i}$; $G_{2',i}$; $G_{3',i}$ are the genetic values of the populations observed in French experiments (DUBIS, 1979).

$\widehat{G_{1',ij}}$; $\widehat{G_{2',ij}}$ are the BLUP indices for the individual bull n°ij.

As expected from economic studies (DUBIS, 1979), only a few pure Friesian bulls are able to compete with Holstein bulls.

Future Research on Progeny Testing

Many results have to be confirmed, especially on the basis of data from other trials in progress, and the connections with

Table 3 Optimum values of coefficient "a" for predicting the aggregate genotype H where $\hat{H} = \hat{G}_1M_1 + a\,\hat{G}_2M_2$, for 324 parameter combinations.

($\hat{G}_1,\hat{G}_2$ = estimated genetic values for milk production and value at 8 days;

M_1,M_2 = corresponding unit margins).

	Parameter value	Number of correlations	Average value of "a"	Standard deviation of "a"
Genetic correlation between milk production and commercial value at 8 days	-0.30	108	0.41	0.10
	0	108	0.60	0.10
	0.30	108	0.79	0.11
Genetic correlation between milk production and value at culling	-0.30	108	0.62	0.20
	0	108	0.60	0.18
	0.80	108	0.59	0.16
Genetic correlation between value at 8 days and value at culling	0	108	0.50	0.15
	0.50	108	0.62	0.16
	0.80	108	0.70	0.17
Repeatibility of the 8 day-value index	0.50	162	0.60	0.17
	0.80	162	0.60	0.19
Heritability of the weight at culling	0.10	162	0.57	0.17
	0.30	162	0.63	0.19
Margin per kg weight at culling (F)	2	108	0.56	0.17
	3	108	0.60	0.18
	4	108	0.64	0.20

performance testing must be estimated. Meanwhile, emphasis is laid on the latter method by improving its generalisation and efficiency within the French dairy selection units.

References

Colleau, J.J., 1978. "Holsteinisation" de la population pie-noire française (prévision de la fréquence des gènes Holstein dans les 20 années à venir).
Bull. tech. CRZV Theix-INRA, 34, 45-50.

Colleau, J.J., Boulanger, P., Dupont, M., Malafosse, A., De Rochambeau, H., 1980. Aperçu sur la variabilité génétique de quelques races bovines françaises.
Bull. tech., Ing. Serv. Agric. n°s 351-352, 467-482.

Cunningham, E.P., McClintock, A.E., 1974. Selection in dual purpose cattle populations : effects of beef crossing and cow replacement rates.
Ann. Genet. Sel. anim., 6(2), 227-239.

Dubis, J.M., 1979. Quelles vaches laitières pour demain? Comparaison de différents types.
Etude ITEB (n° 79063 - 109 pp).

DISCUSSION ON DR. J.J. COLLEAU's PAPER

Chairman: Prof. A. Neimann-Sorensen

Andersen: You have a mixture of Holstein and your own strain of Friesian as we have in Denmark, and we think it is reasonable to estimate parameters in such a population, as in ten to fifteen years time the situation will still be the same. Could you please explain on what the "selling classification" is based ?

Colleau: It is based on the muscularity of the calf.

Langholz: How representative is the sampling of the progeny groups ? Is there any danger of pre-selection in that farmers sell only the poorer calves ?

Colleau: In Trial I, while no accurate figures are available, I am not worried by such a possible bias, because Friesian breeders are not interested in keeping calves themselves. In Trial II, all calves were bought by contract.

Liboriussen: In Table 2 you refer to the trait carcass compacity; what is meant by this ?

Colleau: It is a simple index, the ratio of carcass weight over carcass length. It is therefore highly correlated to carcass weight.

Jansen: Could you adjust for genotype of the cows and so consider it to be one homogenous population.

Colleau: I could do this, however, in the long-run, ten years or so, the French population is going to be essentially pure Holstein in any case.

Foulley: (a) To apply your progeny test under field conditions, how do you account for variation in age at selling ?

(b) You use a selling weight close to birth weight; don't you think it might be dangerous to select on this trait in view of possible calving difficulties ?

Colleau: (a) In the present study, no allowance was made for likely differences between progeny for age at selling. In a new version of the index procedure, a correction for this factor will be made. Furthermore, the index will also be improved by correcting for the genetic type of the mothers mated to the bulls. Finally, the grid will be extended since a significant number of the calves were outside the 1 - 9 classification grid. The extension will be made by estimating the value of an artificial class "zero". This will allow more accurate comparisons between genetic types.

(b) I am not very worried by this problem because of the moderate correlation between selling weight and overall economic weight, in terms of bull index, and therefore it is unlikely that a high selection pressure will be exercised in favour of selling weight. Furthermore, the selling grid gives more importance to classification than to weight. That is not to say that no calving problem will be encountered in the future in the French Friesian. It will, however, be mainly the consequence of the present orientation towards the Holsteins, and not the consequence of a moderate selection on selling weight within the Holsteins by applying the progeny testing system about which I have spoken.

Jansen: (a) What is your criterion for determining the optimum coefficient "a" in Table 3 ?

(b) In other words, when you took reasonable parameters and averaged them, was this optimum ?

Colleau: (a) I used experimental evidence and also "common sense estimations" for choosing likely values of the parameters in calculating value of "a". Pooled experimental evidence was used in estimating the genetic correlation between "value at eight days" and "milk production". A common sense estimation of .05 was used as the correlation between "value at eight days" and "value at culling". Unfortunately, you cannot find all the genetic information you need in the literature, especially for culled cows. Subsequently, values above and below these first estimates were taken in order to determine the sensitivity of the estimation of "a".(b) Yes.

COMPARISON OF DANISH FRIESIAN, HOLSTEIN FRIESIAN AND RED DANISH FOR BEEF PRODUCTION

B. Bech Andersen

National Institute of Animal Science,
Rolighedsvej 25, 1958 Copenhagen V, Denmark.

Summary

The breed comparisons are based on results from grading up experiments, progeny tests for beef production and performance tests of potential breeding bulls.

Holstein Friesian import into the Danish Friesian has led to a slightly positive effect on growth rate, but also to an important negative effect on dressing percentage and muscularity.

In grading up experiments imported semen from the following breeds were tested in crossbreeding with Red Danish:

Finnish Ayrshire, Swedish Red and White, Dutch Red and White, Red Canadian Holstein-Friesian and American Brown Swiss.

The best beef production results were obtained with Dutch Red and White and American Brown Swiss.

Introduction

The dairy and beef production in Denmark is mainly based on dual purpose breeds as SDM (Black and White Danish Cattle), RDM (Red Danish Cattle), and DRK (Red and White Danish Cattle). The Red Danish was the most numerous breed until 1960, when it began to be replaced by SDM, and to some extent also by the Hersey breed. The total number of cattle and the distribution of breeds in 1960, 1970 and 1980 are shown in Table 1.

The various categories of cattle for slaughter are shown in Table 2. Most of the beef production comes from culled cows and young bulls slaughtered at approximately 400 kg live weight. The production of steers is very low. Very few new born calves are slaughtered, which indicates that nearly all bull calves (even Jersey bull calves) are used for beef production.

From the total production of beef, 60% is used for export and 40% for home market. Nearly 60% of the total amount for export is sold to Italy as intact young bull carcasses.

Table 1 Number of cattle in Denmark and the distribution[1] on breeds.

		1960	1970	1980
No of cattle ('000 head)		3,397	2,842	2,958
No of cows ('000 head)		1,438	1,153	1,098
Red Danish cattle	(%)	61.1	35.0	18.7
Black and White Danish Cattle	(%)	18.1	45.5	62.9
Danish Jersey	(%)	15.1	14.9	13.2
Red and White Danish Cattle	(%)	0.9	1.2	3.6
Other breeds	(%)	4.8	3.4	1.6

[1]Based on the statistics from AI centres.

Table 2 Total beef production in Denmark, 1980

	No. of animals		Beef production	
	('000 head)	(%)	('000 tons)	(%)
Culled cows	386	36.2	113	43.0
Heifers	112	10.5	25	9.5
Bulls	504	47.3	112	42.6
Others	64	6.0	13	4.9

On average 66% of the income of cattle farmers is based on dairy products and 34% on beef. Consequently, the beef production capacity of the various dairy and dual purpose breeds used in Denmark is of great importance and also included in our breeding program (Andersen, 1981). For several years the National Institute of Animal Science and the Meat Research Institute have been greatly involved in the testing of growth capacity, feed efficiency, carcass quality and meat quality of various dairy and dual purpose breeds in Europe and North America. The breed comparisons presented in the following are based on results from (1) grading up experiments, (2) progeny tests for beef production and (3) performance tests of potential breeding bulls.

Experimental Design

Import of genes to Red Danish Cattle and Black and White Danish Cattle

In 1973 an experiment was started with the importation of genes from the breeds SRB (Swedish Red and White), FA (Finnish Ayrshire) and MRI (Dutch Red and White). Semen from six highly selected sires of each breed was imported and used on RDM cows. F_1-bull calves were tested for beef production on the test station "Egtved". The experiment was continued with a controlled import of semen from the ABS (American Brown Swiss) and RCHF (Red Canadian Holstein-Friesian) breeds.

In an extensive experiment at the National Institute of Animal Science, Holstein-Friesian young bulls are tested for growth, tissue development and appetite to roughage and concentrates. The experiment is based on serial slaughtering and *ad lib.* feeding with various energy concentrations in the food.

The import of Holstein Friesian genes to SDM is administrated mainly by the National Breed Association, and a large sample of the crossbred progeny has been tested in our progeny and performance test stations.

Progeny test for beef production

is carried out in the central testing station "Egtved". Every year 30 sires are selected for testing, and 10 bull calves from each sire begin the test at 28 days of age. From 1967 to 1972 the calves were fed milk, concentrate and hay according to age. Since 1972 the concentrates have been fed *ad lib.* from a feed dispenser. Feed consumption is individually recorded. The calves are slaughtered at a live weight of 340 kg. The carcasses are dissected partially and muscle samples collected for examination of meat quality.

Performance tests of potential breeding bulls

is carried out in four testing stations. The test period is from 6 weeks to 11 months. Until 1980 the bulls were fed milk, concentrate, sugar beet and hay according to age. The feed consumption is individually recorded. The bulls are weighed every four weeks, and at an age of 10 and 11 months the area of M. longissimus dorsi is measured with ultrasonic equipment.

Results

The beef production results of the first part of the "grading up" experiment with RDM is presented in Table 3. All breed differences are statistically significant, and it is concluded from the experiment, that only MRI can improve the overall beef production capacity of RDM.

Preliminary results from the Holstein-Friesian young bull experiment are shown in Table 4. The average proportion of H-F genes in the tested bulls is approximately 50%. The results indicate that H-F with increasing weight has a faster improvement rate in dressing percentage and muscularity than SDM. However, the increase in deposition of subcutaneous fat in H-F seems higher than preferred. In the table, carcass results of purebred SDM young bulls slaughtered at 470 kg are included for comparison. These results have been obtained in experiments with the same intensity of feeding.

The results of a breed comparison based on 14 years progeny

Table 3 Beef production capacity of F_1 crosses compared with purebred bull calves of RDM (Hansen et al., 1976)

Breed of progeny	No of calves	Least square means for					
		Daily gain (g)	Scand.f.u. kg gain	Dressing percentage	Grading conform.	% fat in meat	Meat tenderness*
RDM	52	1254	3.45	52.7	5.4	1.3	8.5
SRB x RDM	58	1288	3.42	52.0	5.5	1.4	10.4
FA x RDM	55	1252	3.43	51.2	4.9	1.3	9.5
MRI x RDM	57	1293	3.37	53.0	5.8	1.2	10.5

* Shear force values (low values best).

Table 4 Beef production capacity of Holstein-Friesian F_1 young bulls

Breed of progeny	No of calves	Least square means for					
		Daily gain (g)	Dressing percentage	Grading conform.	Lean/ bone	Lean/ fat	% fat in carcass
HF x SDM (340 kg)	47	1273	50.6	5.5	3.84	5.69	12.4
HF x SDM (470 kg)	47	1233	54.0	6.2	4.05	4.33	15.9
(SDM) (470 kg)	(382)	-	(55.9)	(8.5)	(4.15)	(4.60)	(14.9)
HF x SDM (600 kg)	45	1196	55.8	7.1	4.36	3.66	18.6

Table 5 Growth capacity, appetite and feed efficiency of various breeds and breed combinations (data from Danish progeny test stations).

Breeds of progeny	No.of animals	Least square means for: Final weight (kg)	Daily gain (g)	Scand. f.u. day	Scand. f.u. kg gain
From 1967 to 1972 (feeding according to age)					
RDM	1326	373	1098	-	3.71
SDM	883	374	1131	-	3.58
DRK	121	374	1143	-	3.46
From 1973 to 1980 (feeding ad lib.)					
RDM	542	324	1266	4.48	3.56
1/2 RDM x 1/2 ABS[2])	38	324	1321	4.51	3.43
3/4 RDM x 1/4 ABS[1])	18	324	1298	4.32	3.35
1/2 RDM x 1/2 RCHF[2])	61	326	1315	4.59	3.50
3/4 RDM x 1/4 RCHF[1])	20	326	1288	4.41	3.43
SDM	359	324	1293	4.45	3.46
SDM x HF (10-30% HF)[1]	88	324	1319	4.48	3.41
SDM x HF (min.50% HF)[2]	299	325	1327	4.61	3.49
DRK	93	325	1290	4.28	3.33

[1] Sires born and selected in Denmark.

[2] Semen imported.

Table 6 Carcass and meat quality of various breeds and breed combinations (data from Danish progeny test stations).

	Least square means for						
Breed of progeny	Dressing percentage	Grading conform.	Lean/ bone	Lean/ fat	% kidney fat	% fat in meat	Meat tenderness[3]
From 1967 to 1972 (feeding according to age)							
RDM	53.9	5.8	3.8	5.8	-	1.4	11.0
SDM	54.8	8.1	3.8	6.3	-	1.2	12.6
DRK	55.7	8.9	4.0	6.6	-	1.1	13.7
From 1973 to 1980 (feeding ad lib.)							
RDM	52.2	5.4	3.82	5.4	1.00	1.5	7.9
1/2 RDM x 1/2 ABS[2]	51.9	5.5	3.84	6.1	0.83	1.1	8.7
3/4 RDM x 1/4 ABS[1]	52.5	5.3	3.79	5.4	1.06	1.5	7.4
1/2 RDM x 1/2 RCHF[2]	52.0	5.3	3.78	5.5	1.01	1.3	7.6
3/4 RDM x 1/4 RCHF[1]	52.1	5.4	3.82	5.5	0.91	1.3	7.0
SDM	52.4	7.1	3.79	5.6	0.87	1.2	9.3
SDM x HF (10-30% HF)[1]	52.3	6.6	3.79	5.7	0.87	1.2	7.9
SDM x HF (min. 50% HF)[2]	51.9	6.1	3.75	5.2	0.95	1.3	7.7
DRK	53.5	8.0	4.00	5.9	0.76	1.2	8.7

1 Sires born and selected in Denmark.

2 Semen imported.

3 Shear force values (low figures best).

Table 7 Growth capacity, feed efficiency and ultrasonic muscle area of various breeds and breed combinations (data from Danish performance test stations).

	Least square means for					
Breed of progeny	No. of animals	Weight at 11 mths (kg)	Daily gain (g)	Scand.f.u. kg gain	Height at withers (cm)	Ultrasonic muscle area (cm^2)
RDM	833	409	1179	4.42	121	60.5
1/2 RDM x 1/2 ABS[2]	79	436	1259	4.13	124	58.6
1/2 RDM x 1/2 RCHF[2]	125	430	1237	4.21	125	57.7
SDM	195	430	1243	4.22	122	59.1
SDM x HF (10-30% HF)[1]	193	440	1263	4.16	123	61.1
SDM x HF (min. 50% HF)[2]	807	435	1257	4.19	124	58.9
DRK	91	439	1278	4.06	119	63.1

1 Sires born and selected in Denmark.

2 Semen imported.

test results are presented in Table 5 and Table 6, and a breed comparison based on 8 years' performance test results are presented in Table 7.

Importation of semen from HF, ABS and RCHF has led to an improvement of growth rate, no or very little effect on feed efficiency, with a reduction in dressing percentage and muscularity.

On average the economic effect on beef production is minus 125 Dkr. for SDM x HF young bulls compared with purebred SDM. For the RDM F_1-crosses, the positive effect on growth rate and feed efficiency, and the negative effect on carcass quality has neutralised each other.

Most of the R_1 backcrossing combinations have been better than expected from the results of the F_1-crosses, and it is mainly because of a positive selection among the crossbred F_1 -bulls used for breeding.

References

Andersen, B.B., 1981. Danish Breeding Strategies for Milk and Beef Traits in Dairy and Dual Purpose Cattle Breeds. E.E.C. Seminar, Dublin, 13th-15th April, 1981.

Hansen, M., Andersen, B.B., Kousgaard, K. and Buchter, L. 1976. Import of arveanlaeg til RDM. IV. Kødproduktionsegenskaber hos F_1-kalve. National Institute of Animal Science. Report No. 145. 4 pp.

DISCUSSION ON DR. BECH ANDERSEN'S PAPER

Chairman: Prof. A. Neimann-Sorensen

Fewson: Who made the decision to import American Brown Swiss, and if turns out to be a bad decision, who takes the blame ?

Bech Andersen: We are a democratic country ! Thus, a working group of experts (scientists, representatives of breed associations, etc.) advised the national breeding committee on what to do. The national breeding committee made the actual decision and therefore take the responsibility.

Cunningham: The Red Danish situation is unique in Europe in that the invasion of American genes is being planned in advance and with the target of limiting their level to 25%. The experience in other populations is that once this infusion of genes begins, it is difficult to restrain. How do you propose to hold it to 25% ?

Bech Andersen: The plan is provisional, and I have no doubt that in five year's time, some deviation from it will occur - perhaps to an open synthetic with other Scandinavian breeds. I agree that it will be very difficult to keep to a 25% limit on Brown Swiss genes.

Foulley: You changed your feeding system from restricted to ad lib. in order to select for appetite. Is this because of its importance for milk yield ? Do you not think it is enough to improve appetite by selecting for milk yield ?

Bech Andersen: Yes, it must have more or less the same effect. However, the performance test has an important position in the whole breeding scheme, and it is the only place in the system where you can actually select for appetite; furthermore, it can be done at a very early stage.

Langholz: What is your opinion on including cow size in selection procedures. Should one just let indirect selection responses in size follow direct selection for high milk yield and growth rate, and leave it to individual breeders to practise some sort of contrast selection against over-large size. I would prefer that we should put some objective measure on maximum tolerable size.

Bech Andersen: In selecting for growth rate a 10% increase in average daily gain would result in a 40 kg increase in mature size. This is a relatively small increase and furthermore, such cows would yield a higher return at slaughter. In our country farmers want bigger cows. In my opinion we should concentrate on production and efficiency traits in the short to medium term.

ECONOMIC AND GENETIC OPTIMISATION OF DUAL-PURPOSE BULL TESTING AND SELECTION

E.P. Cunningham and Bianca Moioli

The Agricultural Institute, 19 Sandymount Avenue, Dublin, 4, Ireland.

Summary

The cost/effectiveness of various alternative testing and selection schemes for dual purpose bulls used in artificial insemination was investigated. The costs, returns, population structure and parameters were those most relevant to the Irish Friesian AI bred population of approximately seven hundred thousand cows. The discounted gene flow method was used to quantify the economic benefits from testing and selection in net present value terms. These were set against testing and selection costs, and net benefit/cost ratios were calculated.

The results indicated that the programme currently in operation gives a return of about 50 to 1. Expanded testing and selection for beef traits would be highly cost effective giving, for performance testing, a return of better than 100 to 1. A series of sensitivity tests indicated that the results were relatively stable under reasonable deviation from the basic set of assumptions.

Introduction

In Ireland, as in most countries of Western Europe, the principal means of genetic improvement in the cattle population is the testing and selection applied to bulls entering artificial insemination. In the case of the 1.5 m dairy cows, which are now largely Friesian type, this programme is almost exclusively devoted to the screening of Friesian bulls. Since all surviving calves are reared either for beef production, or for milk production, the overall breeding aim must include aspects of both beef and milk productivity. The testing programme in recent years has included performance testing (i.e. beef testing on the bull himself), beef progeny testing (usually on 20 male progeny) and dairy progeny testing (usually on at least 40 daughters). the number of bulls tested and selected at each of these stages for some recent years is shown in Table 1.

Table 1 Number of Friesian bulls tested and selected, and number of Friesian inseminations in recent years

Year	Beef				Dairy Progeny Test		Number of inseminations
	Performance Test		Progeny Test				
	tested	selected	tested	selected	tested	selected	
1974	9	4	24	7	26	7	520,745
1976	20	8	30	10	34	10	522,224
1978	17	8	10	5	45	13	702,727

In planning the development of a complex testing and selection programme for this population, a number of important questions arise. These include

- What relative emphasis should be put on the different traits in the breeding objective, and in particular on the balance of beef versus dairy traits?
- What is the optimum scale and structure of the testing and selection programme for this objective?
- What is the ratio of economic gains to investment for this programme?
- How sensitive are the predicted genetic and economic results of the programme to alterations in the economic, structural or genetic parameters used?

We have recently completed a study of these questions for the circumstances of the Irish dairy cattle population. Results of that study which have particular relevance to the present seminar relate to the balance of beef and dairy traits in the selection objective, and the relative economic returns to investment in testing and selection for beef and dairy traits. These elements of the results are reported here.

Population Structure

The Irish dairy cattle population consists of about 1.5 m cows, almost all of which are now of Friesian type, and about 50% of which are bred by artificial insemination. The present study concerns only the AI bred section. All calves born are reared either as dairy replacements or for beef. Calving and milk production are highly seasonal, and most milk is produced at pasture, with an average concentrate input of about 500kg per cow. Cows average about 5 lactations, and about 30% of dairy cows are normally crossed to beef bulls.

At present, approximately 40 young bulls are progeny tested each year for dairy traits, and most of them are also progeny tested for beef performance of their steer offspring. Some of them are also screened in a beef performance test. Approximately 1 in 4 bulls is selected after progeny testing for widespread use. Once accepted for widespread use, bulls average a lifetime total of about 70,000 inseminations spread, on average, over 5 seasons. The general structure of the testing and selection scheme is shown in Fig. 1.

FIG 1: Structure of testing, selection and utilisation of Friesian Bulls

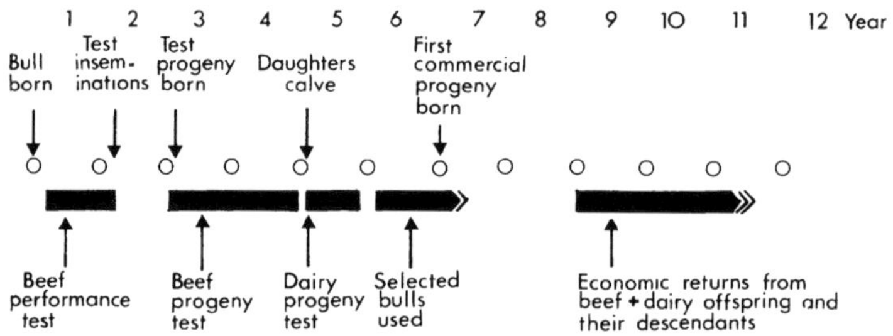

Defining the Breeding Objective

Irish dairy producers are at present paid on a butterfat basis. We have therefore taken the lactation yield of butterfat as the main index of economic output from a dairy cow. The value of a beef animal is a function both of quantity, as measured by the growth potential, and quality, for which the main determinant appears to be carcass conformation. While other characters on both the dairy and beef side (e.g. milk speed, udder conformation, calving ease, feed efficiency, meat quality) would need to be taken into account in ultimate bull selection, most of the selection decisions are normally based on these three main production criteria.

Economic analyses have shown these three characteristics to have net economic values to the producer as follows:

Growth rate	-	44p per kg liveweight
Conformation score	-	150p per point on a seven point carcass conformation scale

Butterfat - 140p per kg

The aim of the selection programme is to select those bulls which have maximum genetic value for an appropriately weighted combination of these three traits. In the Irish population most young bulls which go on test are purchased from Britain. The genetic benefits from bull selection are therefore, at present, limited to those passing in commercial inseminations in the bull - cow path. This means that bulls should be selected for the balance of traits as these traits are expressed through a commercial insemination. The Discounted Gene-Flow Method (McClintock and Cunningham, 1974) was used to express the total genetic consequences of commercial inseminations in net present value terms. Effectively, this means that each trait in the selection objective is weighted not just by its economic value, but also by a value representing its expression in the population, as a result of a single insemination. These "Discounted Expression" factors are functions of the percent of cows crossed to beef bulls, the number of lactations per cow, the net fertility and calf survival rates in the population, and the financial discount rate used to evaluate future gains. For the values appropriate to the Irish population (5 lactations per cow, 30% beef crossing, 85% net calf crop, and 8% discount rate) the number of discounted progeny equivalents for a beef trait is 0.77 and for a dairy trait is 1.14.

Multiplying the three target traits by their relative economic values, and by these discounted expression factors gives the selection goal:

$$(0.77)\ (44)\ \underline{GROWTH} + (0.77(150)\ \underline{CONFORMATION} + (1.14)\ (140)\ \underline{BUTTERFAT}$$

Using this selection goal means that we are aiming to maximise the net present economic value of the total genetic merit conferred on the population with each insemination.

Genetic and Phenotypic Parameters

Analyses of recent historical data have provided us with

estimates of the heritabilities, genetic and phenotypic correlations, and variances for the population. These are shown in Table 2

Table 2 Genetic and phenotypic parameters. Heritabilities on diagonal, genetic correlations above, phenotypic below.

	Growth	Conformation	Butterfat	Mean	CV%
Growth	.30	.60	.0	500 kg	10
Conformation	.30	.23	-.30	4.2 points	19
Butterfat	0	-.30	.20	108 kg	18.5

Predicted Genetic Gains

When test information is available on bulls from their own performance test, from the beef progeny test of their steer offspring, and from the dairy progeny test of their daughters, several different kinds of selection are possible. Selection on the following bases is considered:

1. Beef performance test
2. Beef progeny test
3. Dairy progeny test
4. Beef and Dairy progeny tests
5. Dairy progeny test following prior selection on beef performance test
6. Beef and Dairy progeny test following prior selection on beef performance test.

For cases where selection is carried out in a single stage, the predicted gain is

$$\Delta_g = D\sigma$$

where D is the selection differential and σ is the standard deviation of the index used.

For selection in two stages, the projected genetic gain was taken as the sum of the gains from two stages of selection:

$$\Delta_g = D_1\sigma_1 + D_2\sigma_2$$

FIGURE 2.

Discounted net benefits per bull selected vs costs per bull selected

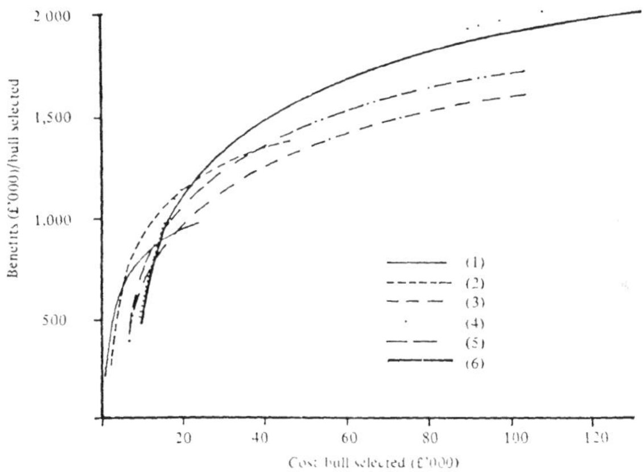

Discounted benefit/cost ratio vs cost per bull selected

Benefit/cost relationships for increasing investment in bull testing and selection for six alternative programmes.

1) Beef performance test.
2) Beef progeny test.
3) Dairy progeny test.
4) Beef & Dairy progeny test.
5) Dairy progeny test after 50% selection on Beef performance test.
6) Beef & Dairy progeny test after 50% selection on Beef performance test.

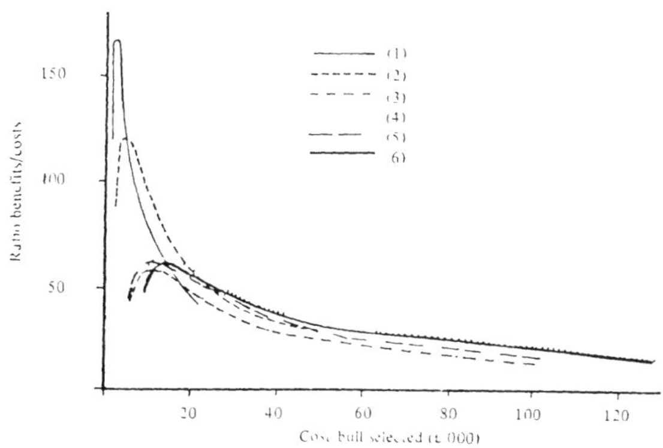

where D_1 and D_2 are selection differentials corresponding to the proportions of those bulls tested which are selected at Stage One, and Stage Two, and where the index at the second stage has been calculated from parameters adjusted to take account of the effect on variances and co-variances of whatever selection has taken place at the first stage (Cunningham, 1975).

All calculations were carried out using the computer programme SELIND (Cunningham, 1970).

Testing Costs

The costs of the different tests were estimated from current experience, and the cost per bull tested in each case was as follows:

Beef Performance Test	£1,101	
Beef Progeny Test	£2,154	
Dairy Progeny Test	£4,182	(if beef progeny test also carried out)
Dairy Progeny Test	£5,127	(if no beef progeny test)

These costs are all discounted forward from the various stages in the bull's life at which they are incurred to the year after the progeny test is complete. This is also the year to which the benefits arising from commercial inseminations are discounted. Costs and benefits have therefore both been discounted to the same point in time.

Economic Evaluation

Numerous alternatives of the basic programme were explored, but the main results are summarised in Fig. 2. The horizontal scale in both of the graphs in Fig. 2 is the same, and represents the investment per bull ultimately selected. This in turn depends on the number of bulls tested for each selected, and on the unit cost for the particular kind of testing involved. A maximum selection intensity of one in twenty was specified. Thus, the maximum expenditure per bull selected on performance test only, where the testing cost per bull is approximately £1,000 amounts to £20,000. On the other hand, it would be possible to spend £100,000 per bull selected on dairy progeny test, if one tested twenty for each selected. The vertical scale on the first graph represents the net discounted benefits to the population per bull selected. This in turn, is simply the net value per commercial insemination, scaled up by the 70,000 life-time inseminations expected from that bull. Thus, the first graph shows the relationship of returns to expenditure, when both are related to the same unit (bull selected) and evaluated at the same point in time. It thus shows the true relationship of benefits to costs.

Perhaps more interesting is the relationship of the ratio benefits/costs to the costs. This is shown in the second graph. This clearly shows that for each of the selection schemes there is an intermediate optimum, which is rapidly attained, and which falls off nearly as rapidly. In all cases, this optimum is in the region of a one in four selection. It is interesting that the ratio of return to investment exceeds 150 to 1 for performance testing, is better than 100 to 1 for beef progeny testing, and exceeds 50 to 1 for all programmes involving dairy progeny testing. These are quite remarkable returns to investment by any standards. In our circumstances, they understate the situation because the same investment also gives benefits a generation later in the non-AI half of the population, through the use of sons of selected bulls. Furthermore, the genetic gains flowing through the sire-son path have not been taken into account.

Two-Stage Selection

Two of the testing and selection alternatives involved two stages. In each case, the first stage was a beef performance test. The balance of selection intensities in the two stages was examined. In general, if the overall selection intensity was held fixed, then overall genetic gain fell off quite sharply if more than 50% selection was exercised on the first stage. If overall selection intensity were not limited, then the effect of increasing intensity on the first stage was simply to reduce the variance, and hence the selection response at the second stage for the two beef traits. In addition, there was a slight negative correlated response in butterfat because of the negative genetic correlation with beef conformation. However, even at very high levels of selection intensity on performance test, this effect in no case reduced the total gains by more than 5%.

Each of the two-stage options involves a combination of a beef performance test, beef progeny test, and dairy progeny test. The cost per bull tested for each of these components is known. In addition, for any particular combination, and for specified selection intensities at the first and second stages, we can calculate a measure of the contribution of each component to the overall gains. This measure is the % reduction in overall gain if that particular component is omitted. Since we can calculate the overall annual benefits in the population from a particular testing and selection combination, we can estimate the benefits from inclusion of each component in the selection programme. This benefit can then be related to the testing cost involved to give a benefit/cost ratio for that particular component.

Table 3 Benefit/cost ratios for components of the testing programme. (10% overall selection, 50% selection at Stage 1).

Testing and Selection Combinations	Beef Performance Test	Beef Progeny Test	Dairy Progeny Test
Beef & Dairy Progeny Test after beef performance test	21	12	27
Dairy progeny test after beef performance test	33	-	28

Table 3 shows these ratios for each of the three testing and selection combinations considered, and assuming a 10% overall selection with a 50% selection after the first, or performance test, stage.

The first thing to note is that these ratios are considerably lower than those given for overall cost-effectiveness at the same overall selection intensity (Fig. 2). The reason for this is that because of correlations between the variates measured, when any one of them is omitted its contribution is to some extent taken care of by those which remain.

The overall pattern is that for the alternatives where a single stage of beef testing is combined with the dairy progeny test, the cost-effectiveness of both the beef and dairy tests are similar. If beef testing is carried out both through a performance test and a progeny test, then at these selection intensities the beef performance test gives a considerably better return on expenditure.

Components of Response to Selection

For any particular selection combination, set of parameters, and intensities, the total economic response to selection is due to genetic change in the three target traits. It is possible to calculate the % contribution of gain in each trait, as a response to each testing input, in this aggregate gain. For the two-stage testing combinations, at a specified set of selection intensities, this breakdown is given in Table 4.

Table 4 Distribution of overall genetic gain over the three traits (overall selection intensity 10%; 50% selection in Stage 1).

Testing Combination	Percent of total gain due to gains in				
	Growth from		Conformation from		
	Performance test	Progeny test	Perform-ance test	Progeny test	Butter-fat
Beef & Dairy Progeny Test after beef performance test	17	22	5	3	53
Dairy progeny test after beef performance test	27	-	3	-	71

Sensitivity Testing

Genetic Correlations: Increasing the genetic correlation between growth and butterfat from 0 to +0.2 increased the net gains by 11%. Decreasing the genetic correlation of growth and conformation from +0.6 to +0.3 reduced the net gains by 5%. The only unfavourable correlation in the basic parameter set was that between conformation and butterfat, which was estimated to be -0.3. When this was reduced to 0, the net effect was an increase in net gains of 7%. When, in addition, the economic value of conformation was increased by 50% the effect was an increase of net gains of 14% over the basic parameter set.

Economic Weights: Changes in the relative economic weights of the three traits are mostly likely to concern the relative weighting of butterfat to the two beef traits. Runs were done in which the economic weighting of butterfat was increased by 25%. and also decreased by 25%. In the first case, an increase in net gains of 16% was obtained, while in the second, a decrease of 14% followed. This range of ± 25% covers more than the variation experienced over the last decade in the ratio of beef/ milk prices.

Scale of Milk Production: The average level of milk production in Ireland is at present approximately 3,000 kg, though tending to increase. With an increase in the mean production, the phenotypic variance will also increase. We therefore did a further run in which the mean milk production is assumed to be 4,500 kg, and in which the standard deviation in butterfat per cow is assumed to go up in proportion i.e. by 50%. The effect of this change was to produce an increase of 33% in the net gains.

Restricting Change in Beef Conformation: For some combinations of parameters, selection of an optimal economic balance of traits caused a slight reduction in mean beef conformation score. It might therefore be reasonable in some circumstances to impose a restriction that selection be optimised while maintaining mean conformation score. The effect of imposing this restriction for the basic parameter set is to cause a reduction in net gains of 4%

Discount Rates: Variation in the discount rate has been shown to affect the net present value of genetic gains by about 5% increase or decrease per 1% change in the discount rates. An increasing discount rate will also increase the net present value of testing costs. Changes for the discount rate therefore have a double effect on cost/benefit ratios. The net effect is approximately a 9% increase or decrease in the benefit/cost ratio per 1% change in the discount rate.

The present programme (assuming an 8% discount rate) is calculated to have a benefit/cost ratio of 53. At a discount rate of 10%, this ratio would drop to 43, while at a discount rate of 6%, it would increase to approximately 63. At a "social" discount rate of 3%, which, it can be argued, should be the appropriate rate for investments of this kind (Smith, 1978) the benefit/cost ratio would increase to 85. If discounting is disregarded, i.e. a discount rate of zero is used, the ratio of benefits to costs increases to 119.

Conclusions

The general conclusion from these additional runs is that any likely deviation from the basic parameter set will probably give results within 10% of those obtained with the basic set. Such a deviation would not significantly alter the balance of contribution or gain from the different traits. The general conclusions regarding cost/benefit ratios from alternative programmes are therefore likely to remain broadly correct under foreseeable variations from the basic parameter set used.

The pattern of use of selected bulls is also not critical, though the higher the discount rate, the more it pays to use bulls heavily as soon as their testing is complete. Because of their multiplicative effect on cost-effectiveness control of testing costs can be very important. The discount rate used is also important, since the difference in cost-effectiveness ratios calculated using a social rate of 3% and an actual bank rate of 15% is about 300%.

The results in general suggest that all aspects of testing

of bulls for artificial insemination in Irish circumstances are highly cost effective and that initial screening of bulls for beef traits in a performance test is the most effective of the three testing stages examined. Any of the selection programmes will improve growth and butterfat production. In some cases, beef conformation might marginally decline. However, the present level of this trait can at least be preserved at small cost by minor adjustment of the selection programme.

References

McClintock, A.E. and Cunningham, E.P. Selection in dual-purpose cattle populations: defining the breeding objective.
Animal Production, 18, : 237, 1974.

Cunningham, E.P. SELIND - computer programme specification; 1970.

Cunningham, E.P. Multi-stage index selection.
Theoretical and Applied Genetics, 46 : 55, 1975.

Smith, C. The effect of inflation and form of investment on the estimated value of genetic improvement in farm livestock.
Journ. Anim. Prod. 1978, 26 : 101-110.

DISCUSSION ON PROF. E.P. CUNNINGHAM'S PAPER

Chairman: Prof. A. Neimann-Sorensen

Thiessen: Could you explain what you mean by "per insemination" ?

Cunningham: One can define that in a number of ways, but what we mean by it is as follows : per insemination plus up to two repeats and including 85% survival rate.

Jansen: Are the costs of milk recording at farm level included ?

Cunningham : Yes.

Fewson: You only considered the selection of bulls; would the inclusion of selection of bull-dams influence the conclusions ?

Cunningham : There are four possible paths that could be considered. However, since in Ireland bull-dams are not often selected (many young bulls going on test are in fact imported from the UK) they were not included in the study. Since other paths are not considered, the benefits given are greatly understated. For example, many cows are bred to sons of selected bulls and this effect is not included in the benefits.

Bech Andersen: (a) Your present selection intensity is quite limited. Do you recommend increasing it ?

(b) Are your recommendations to be taken up by the AI centres ?

Cunningham: (a) Our recommendation is that we should double the number of bulls progeny tested and performance tested.

(b) Yes, I think so. The problem is that we have eight AI centres and each makes its own decision. However, I think they are persuaded by the arguments.

Foulley: (a) How do you calculate profitability ?

(b) How dependent are your results on the beef crossing rate ?

Cunningham: (a) Profitability is the ratio of discounted benefits to costs.

(b) I don't know how a change in the beef crossing rate will affect the overall returns on investment, but it will change the proportion of benefits arising from beef or milk.

Jansen: What is the net present value of an insemination ?

Cunningham: About £20, and the cost is £7, so the return on investment is approximately three to one.

THE ECONOMIC COMPARISON BETWEEN THE HOLSTEIN AND NORMAN BREEDS

J.J. Colleau and G. Lienard*

Station de Génétique Quantitative et Appliquée
INRA - 78350
Jouy-en-Josas
France

*Laboratoire d'Economie de l'Elevage
INRA - 63110
Theix
France

Summary

For many years, dual purpose breeds were the selection objective in Western Europe, especially in France. This has been changing in recent years with the continuous grading up of the Friesian with Holstein genes.

The latter breed may destabilise the national cattle production in the long term: too strong a movement toward the Holstein may lead either to milk surpluses which will be difficult to market, or to a dramatic decrease in the number of cattle and a corresponding decrease in beef production. In these circumstances, dual purpose breeds still have a role to play, provided they are not being progressively given up by the dairy producer because of their lower profitability.

This paper attempts to estimate the profitability of the Holstein compared to the Norman breed based on figures produced from an experiment designed for that purpose. These results confirm the difficulty for a breed like the Norman, whose milk production is some 1700 kg less than the Holstein, of competing, in spite of its better milk composition, beef production and reproduction. The major items where it might compete better with the Holstein are:-

- payment for milk according to its processing value, giving special weight to protein yield; replacing the present payment system by one based on payment for valuable matter would have the effect of reducing the economic difference between the breeds by the equivalent of 500 kg per lactation.
- selection for milk production in dual purpose breeds, while maintaining their present genetic level for beef production, must also be steered.
- finally the dual purpose breeds are more adaptable to the severe environmental conditions in certain areas of France. The Holstein would be quite unfit for some of these areas.

Introduction

The introduction and development in France of the specialised dairy strains such as the Holstein gives rise to a twofold problem : first, that of their economic advantage at the national level, expressed in terms of the balance between milk and meat production, and secondly that of their economic advantage for the farmers, expressed in terms of profitability.

That introduction mainly affects the French Friesian breed in which the Holstein crossbreeding is made. In that field, we have some results, either of a zootechnical nature (JOURNET et. al., 1973, CHEVALDONNE, 1978; REGIS, 1979-1980) or of a microeconomic nature (GUITTET, 1975; DUBIS, 1979) allowing us to think that introducing the Holstein breed within the Friesian breed is economically positive, despite its well-known weaknesses, especially for producing meat. Of course, the result depends on the management conditions, especially on the quality of the fooders used for food : the advantage of the Holstein is less clear when it is impossible to use maizesilage for winter food. However, this set of results have little chances to question again the present trend within the breed. Indeed, a theoretical study realised from the genetic composition of the testing batches and of A.I. service bulls allows to foresee that, within 10 years, the French Friesian population will be heavily influenced by the Holstein genes (COLLEAU, 1978a).

Because of this present evolution of the French Friesian, the breeders using the dual-purpose breeds, especially the Norman and Montbeliarde breeds (about 35% of the french dairy cows) may choose between staying in the first breed or opting for a Black-pied cow graded-up with Holstein, or even for a pure Holstein cow. To what extent and in what conditions is it economically interesting to make that change? This question concerns a genotype-environment interaction, the analysed character being not a precise zootechnical character but the overall economic profitability. This is a question concerning the national policy for cattle, as well. Indeed, it would be pertinent to discourage the Holstein crossbreeding on dual-purpose breeds in the conditions where it brings nothing valuable to the breeder whereas it brings big risks of disturbing the balance of the national commercial exchanges of meat and milk.

To give the right answer to this fundamental question is a very difficult exercise, in our opinion, for the following essential reasons:

- it would be necessary to know all the zootechnical yields (milk, meat, reproduction, health state...) of the genetic types being compared i.e. Holstein vs dual-purpose breeds, in a range of environments, described themselves by means of quantitative and qualitative variates.
- it would be necessary to scan a broad range of economic conditions in reference not only to market prices but also, and above all, more or less severe financial constraints in the farm.

An experimentation has been set up since 1968 by the INRA, in the Le Pin-au-Haras farm (Normandy) to directly compare the Holstein and Norman herds. The Holstein was then not frequent in France, at the beginning. It has been aiming at gathering technical criteria for setting up the economic comparison between these breeds, but in a rather narrow range of managemental conditions. From now on, it is developing towards a study of the genotype-environment involving the same breeds, in order to give more pertinent answers to the questions coming from the professionals and the administration (Table 1).

The first experimentation will come to an end with the slaughter of the last cows, in 1985. However, the necessity of giving a preliminary result was urged : this first work was published in French two years ago (COLLEAU, LIENARD and JOURNET, 1979) and we intend in this paper to describe its general lines. Later on, a definitive result should be published, from more strongly established experimental results and from an improved methodology, taking account of all the criticisms or suggestions made at the present seminar.

Table 1. Experiments on cattle at the Le Pin INRA farm (about 160 lactating cows).

	Genetic types compared	Management system
Old experiment AI (AI 1968→1978)	Holstein selected (HF+) Norman selected (NO+) Criss-cross Holstein-Norman Milk divergent selection in Norman (NO^{-})	Semi-intensive cow feeding Comparison on 3 lactations culling after 3 lactations Young bull production at first, then (young bull production + veal production).
New experiment AI (AI 1979→1985)	Holstein highly selected on valuable matter yield (HF^{++}) Control Holstein (HF T) (the same 6 average Holstein bulls) Norman highly selected on valuable matter yield (NO^{++})	Two management systems intensive: - good winter food (→16kg milk) - silage help if grass shortages at summer - liberal allocation of concentrate - good protein feeding after calving - good calving season by means of heat induction Extensive: - the contrary for all these factors Common: - comparison on 5 lactations - culling after 5 lactations - veal production

I. Basic Model

a - Difficulties and limits of modelling

Two preliminary questions arise:

- In our experimental conditions, have we at our disposition all the data allowing us to set up a pertinent economic evaluation?
- Are the experimental conditions representative or more exactly, how do we get a reasonable idea of the results that might be obtained in other situations ?

Even in very well monitored experimental conditions, it is not possible to observe all the variates necessary to establish the total result, involving the incomes, the variable costs, the surfaces and the related fixed costs. Some values could be observed directly within the experimentation such as milk yields or carcass weights, other values needed to be estimated and built up, such as the commercial value of the cows at culling or the necessary surfaces for maintaining different categories of cattle. All the hypotheses adopted will appear when we shall expose the zootechnical and market bases used for the calculations.

Strictly speaking, our experimental conditions are peculiar. However, it seemed that it was possible to analyse rather well some situations not studied in Le Pin, where the variations involved culling rate, mortality rate, the policy of marketing surplus heifers and the policy of using males. The number of extra hypotheses in comparison to the strict experimental data appear to be small, if not zero.

Other situations seem *a priori* to be very interesting to analyse. They involve for example the energetic level of the basal food intake, the concentrate allocation or the genetic value for milk yield since they might affect the results, giving evidence about the respective utilisation domains of the dairy breeds and dual-purpose breeds. These "different" situations are so far from our conditions that the number of extra hypotheses necessary for "simulating" the data increases rapidly. Nevertheless, we thought that we should make an attempt in that area, at least to give ideas for further experimentations.

b - Model used

The milk yield results were known with a minimal accuracy only on the first and second lactation, at the date of the study (1976-1977). So we imagined a model where we considered only these two lactations and all the incomes, costs, surfaces related to them.

The extrapolation made to a complete herd, where all the lactations are represented, seems to be a rather good approximation, as the differences for milk yield do not seem to depend upon age, and as the age affects are not so high in our conditions (no mass selection for milk yield).

b.1 - Economic indicator

It is necessary to choose an economic indicator having a value as general as possible but close to the farm economics. We will use, not a "margin" per cow, but a "margin" per ha of fodder area : that is indeed the limiting factor for most pertinent comparisons of situations differing not only for the cow performances but also for their progeny's performances and destinations.

Further, it is very difficult to take into account the real differences for consuming grass and to represent them in terms of surfaces. For the animals were managed together at pasture for experimental purposes and in these conditions, no separate breed evaluation could be made during that period. That is the reason why we supposed that the herds were fed with maizesilage only (the concentrate consumptions being those observed during the experimentation) in order to make allowance for the actual needs of the animals, all the year round. The weakness of that simplification is to make the results less representative in their absolute value. But the hierarchy of the results across situations, should not be affected very much by using that process.

So, the "margin" thus established is a "margin" per maize silage ha .

To work it out, we only took into account the elements

varying across situations.

- Incomes (milk and meat)
- Costs : only those involving the herd (concentrate, veterinary, inseminations...).

In return, the expenses common to the different situations (charges for producing and storing maize silage, structural changes) are not taken into account : each farmer may evaluate them from his own conditions.

The necessary amount of maizesilage was calculated from the difference between feed needs and concentrate allocations. The maizesilage amount was then transformed in fresh maize surfaces.

b.2 - Technical and economic bases

The zootechnical observed data are those known at the end of 1976. The prices used are those of the years 1976-1977.

Due to the relative recency of the experimentation at that time, some data had to be estimated.

The total set of the values used, (observed or built up again) is shown in Table 2.

II. - Results in Situations approaching our Experimentation

We parametered 40 situations of that kind that is to say 32 situations of pure breeding and 8 situations of Charolais crossbreeding, the crossbreeding rate being the maximum rate permitted by the replacement needs.

The parameters relating to the 32 situations of pure breeding are: presence or absence of young bulls (2 levels), the destination of the surplus heifers (4 levels), the culling rate (2 levels), calf mortality rate (2 levels).

The parameters involving in the 8 situations of crossbreeding are: presence or absence of young bulls (2 levels), culling rate (2 levels), calf mortality rate (2 levels).

We report in Tables 3,4,5, the final results for each of these 40 situations that is to say the margin obtained by the two breeds and the difference between them (Holstein minus Norman), the results being expressed in French francs and in

Table 2 - Technical and economic bases taken for calculating (figures of the year 1976-1977)

	Source of Income	Holstein breed	Norman breed	Holstein-Norman	Observations
Incomes	Milk production	Milk payment: 0.95 F per kg ± 0.012 F per fat% above or below 37%			Observed in Le Pin experiment
		Averages (L_1+L_2)			
		Milk yield 5 752 Fat% 36.9 Protein % 30.3 Lactation length 338	3 466 41.4 34.3 311	2 286 - 4.5 - 4.0 + 27	
	Culled cows	3 063 F	3 738 F	-675 F	Indirectly built up from Le Pin results
	Culled heifers at 30 mths (10% of the inseminated heifers)	2 888 F	3 644 F	-756 F	built up
	8 days calves	350 F x Charolais 850 F	700 F x Charolais 1 000 F	-350 F -150 F	Estimated (see GUITTET and DUBIS)
	Young bulls (16 months)	2 993 F x Ch 3 509 F	3 324 F x Ch 3 783	-327 F -274 F	Observed
Costs varying according to breed	Concentrate	Cost = 1.20 F/kg			Observed
		Averages (L_1+L_2)			
		1 249 kg	561 kg	688 kg	
	Insemination Cows Heifers	113 F 86 F	53 F 48 F	+ 60 F + 28 F	built up
	Veterinary	200 F/cow and progeny	150 F/cow and progeny	50 F	estimated
Surfaces (maize giving 7310 milk feed units effectively consumed per ha)	Cows (sum L_1+L_2)	0.81 ha	0.75 ha	+0.06 ha	built up
	Heifer till the first calving	0.59 ha	0.59 ha	0	built up
	Young bull till slaughter	0.34 ha x Ch 0.33 ha	0.31 ha x Ch 0.32 ha	+0.03 ha	built up

Table 3 Results of the 32 situations in Purebreeding

First line : Surface (ha)

Second line : Margin/ha in F - (Margin/ha expressed in equivalent kg 37°/milk)

MALE PROGENY MARKETED AT 8 DAYS

Destination of surplus heifers	Culling rate	Calf mortality rate	Holstein breed	Norman breed	Difference Holstein minus Norman
	%	%			
8 days calves for veal production	20	5	1.037 7 683	0.981 6 918	0.056 765 (805)
	20	15	1.037 7 621	0.981 6 789	0.056 832 (876)
	25	5	1.067 7 434	1.013 6 733	0.054 701 (738)
	25	15	1.067 7 376	1.013 6 612	0.054 764 (804)
8 days calves for breeding	20	5	1.037 7.963	0.981 6 918	0.056 1 045 (1100)
	20	15	1.037 7 846	0.981 6 789	0.056 1 057 (1113)
	25	5	1.067 7 638	1.013 6 733	0.054 905 (953)
	25	15	1.067 7 528	1.013 6 612	0.054 916 (964)
In-calf HF 5 000 F NO 4 000 F	20	5	1.311 7 246	1.255 6 289	0.056 957 (1007)
	20	15	1.257 7 262	1.201 6 281	0.056 981 (1033)
	25	5	1.273 7 135	1.219 6 277	0.054 858 (903)
	25	15	1.220 7 149	1.166 6 271	0.054 878 (924)
In-calf HF 4 500 F NO 4 000 F	20	5	1.311 7 114	1.255 6 289	0.056 825 (868)
	20	15	1.257 7 151	1.201 6 281	0.056 870 (916)
	25	5	1.273 7 032	1.219 6 277	0.054 755 (795)
	25	15	1.220 7 070	1.166 6 271	0.054 799 (841)

MALE PROGENY MARKETED AS YOUNG BULLS

Destination of surplus heifers	Culling rate	Calf mortality rate	Holstein breed	Norman breed	Difference Holstein minus Norman
	%	%			
8 days calves for veal production	20	5	1.319 7 231	1.236 6 744	0.083 487 (513)
	20	15	1.289 7 222	1.209 6 656	0.080 566 (596)
	25	5	1.341 7 051	1.262 6 595	0.079 456 (480)
	25	15	1.312 7 037	1.235 6 512	0.077 525 (552)
8 days calves for breeding	20	5	1.319 7 452	1.236 6 744	0.083 708 (745)
	20	15	1.289 7 043	1.209 6 656	0.080 747 (786)
	25	5	1.341 7 214	1.262 6 595	0.079 619 (652)
	25	15	1.312 7 161	1.235 6 512	0.077 649 (682)
In-calf HF 5 000 F NO 4 000 F	20	5	1.593 1.593	1.510 1.510	0.083 697 (734)
	20	15	1.509 6 981	1.429 6 250	0.080 731 (769)
	25	5	1.547 6 856	1.467 6 240	0.080 616 (648)
	25	15	1.465 6 884	1.388 6 236	0.077 648 (682)
In-calf HF 4 500 F NO 4 000 F	20	5	1.593 6 790	1.510 6 253	0.083 537 (565)
	20	15	1.509 6 887	1.429 6 250	0.080 637 (670)
	25	5	1.547 6 772	1.467 6 240	0.080 532 (560)
	25	15	1.465 6 818	1.388 6 236	0.077 613 (645)

Table 4 Results of 8 situations in Crossbreeding

First line : Surface (ha).

Second line : Margin/ha in F - (Margin/ha expressed in equivalent kg 37 % milk)

Destination of male and surplus female progeny	Culling rate	Calf mortality rate	Holstein breed	Norman breed	Difference Holstein minus Norman
Male and surplus female progeny marketed at 8 days for veal	20	5	1.037 8 114	0.981 7 191	0.056 923 (972)
	20	15	1.037 7 966	0.981 7 007	0.056 959 (1009)
	25	5	1.067 7 749	1.013 6 887	0.054 862 (907)
	25	15	1.067 7 609	1.013 6 759	0.054 850 (895)
Male progeny as young bulls, and surplus heifers marketed at 8 days for veal.	20	5	1.315 7 595	1.241 6 988	0.074 607 (639)
	20	15	1.286 7 518	1.213 6 853	0.073 665 (700)
	25	5	1.338 7 323	1.264 6 784	0.074 539 (567)
	25	15	1.310 7 212	1.237 6 655	0.073 557 (586)

milk equivalent (kg). In order to allow comparisons not only according to unit of area but also to unit of herd size, we add the surfaces used for maize: it must be kept in mind that, within our model, these areas are those resulting from the utilisation of one 1st lactation heifer (reference animal) and of all the animals related to her (culled cows L_1, dairy cows L_2, corresponding male and female progeny).

In order to make the comparisons between all these situations easier and to extract from them approximate but clear ideas, we imagined to process breed differences like an analysis of variance: estimation of mean effects for the parametered factors and estimation of the interactions between them.

a - Situation of pure breeding

Across the 32 situations of pure breeding, the mean difference between breeds for the margin per ha corresponds to 778 kg milk at 37°/oo fat. The results in Table 6 show the means obtained for each possibility of each factor (destination of surplus hiefers, presence of young bulls, culling rate, mortality rate).

The breed difference is at its minimum when the surplus heifers are marketed at 8 days as young calves for veal production but the value obtained (671 kg equivalent milk) is nevertheless important. The difference is maximum (874 kg) when the young Holstein heifer calves are sold for breeding. To market in-calf Holstein heifers allows a big difference (838 kg), as well.

To use all the male calves for young bulls makes the breed difference decrease significantly : 639 kg milk, whereas the difference is 915 kg milk when the male calves are marketed at 8 days for veal production. This result is not unexpected since the difference Holstein minus Norman is negative with young bulls, whereas it is positive with cows and heifers.

An increase in the culling rate lowers the difference Holstein-Norman very clearly. The difference declines from 817 kg to 731 kg when the culling rate increases from 20% to 25%. That is the consequence of an increasing proportion of the income coming from the culled cows, which for the

Table 5 Average effect of the variation factors on the difference of margin/ha between the two breeds (expressed in kg 37 ^{o}/oo milk) for the 32 situations of purebreeding

Factors	Mortality	Average	Deviation to the general average (778 kg)
Policy of heifer marketing	8 days (meat)	671	-108
	8 days (breeding)	874	96
	in-calf (prices 1)	838	59
	in-calf (prices 2)	733	- 46
Young bulls	Yes	639	-139
	No	915	+137
Culling rate	20%	817	39
	25%	739	- 39
Calf mortality rate	5%	734	- 24
	15%	802	24

Table 6 Average effect of the variation factors of margin/ha between the two breeds (expressed in kg 37 ^{o}/oo milk) for 16 situations (8 in purebreeding - 8 in crossbreeding).

Factors	Mortality	Average	Deviation to the general average (729 kg)
Crossbreeding with Charolais	Yes	788	+ 59
	No	671	- 59
Young bulls	Yes	583	- 146
	No	876	+ 146
Culling rate	20%	764	34
	25%	695	- 34
Mortality rate	5%	703	- 27
	15%	756	27

difference Holstein-Norman is negative.

The interaction effects between the main factors analysed were calculated . These values can sometimes be important in absolute value and comparable to some main effects. Reasoning on 32 values, 5 are smaller or equal to -30 kg and 4 greater or equal to + 30 kg. These findings show that the preceeding discussion on the quantitative effects of the variation factors allows us only to give rough ideas, as they cannot be, strictly speaking, added together.

b. Comparison between beef crossbreeding and pure breeding

This comparison was established for 8 situations of beef crossbreeding and 8 corresponding situations of pure breeding, where the females are marketed at 8 days (Table 7). It can be seen that the utilisation of beef crossbreeding allows an increase of 108 kg for the difference Holstein-Norman. This result is not surprising since the improvement of the meat commercial value as a result of beef crossbreeding is more important for the Holstein breed than for the Norman breed.

The estimation of the other effects (presence or absence of young bulls, culling rate, mortality) are very similar to those obtained on the overall set of the pure breeding situations.

III. - Results in Situations Differing from our Experimentation

The parameters analysed previously have an influence upon comparative breed profitabilities but it is clear than in any of the 32 situations, the Norman breed succeeded in being equivalent to the Holstein breed.

However these situations do not represent the real range of situations since for example, we did not allow the level of milk yield to vary nor that of feeding (fodder and concentrate) while they are essential items.

a - Genetic level of the Norman cows

The previous results were set up by using average genetic levels for the two breeds and demonstrated the superiority of the

Table 7 Variation of difference of margin per ha between the two breeds according the basal food and the genetic level of the Norman cows

First line : Surface (ha)

Second line : margin per ha in kg equivalent milk

Basal food	Holstein minus original Norman	Holstein minus improved Norman
No 1		
Maize silage 16 kg 4% milk	1,265	609
No 2		
Maize silage 10 kg 4% milk	906	417
No 3		
Hay 4 kg 4% milk	721	259

Fixed parameters:
No beef crossbreeding
Culling rate : 25%
Mortality rate 5%
Male and surplus heifers marketed at 8 days

Table 8 Unfitted concentrate allocation (50% of the theoretical needs)

Basal food	Holstein	Original Norman	Difference of margin/ha kg equivalent milk
Maize silage	5 037 (1)	3 357	
16 kg 4% milk	1 349 (2)	1 156	
	6 535 (3)	6 054	506
Hay	1 169	1 096	
4 kg 4% milk	4 769	4 650	125

(1) Milk yield per lactation

(2) Surface (ha)

(3) Margin/ha (FF).

Holstein breed. So we investigated if it were possible for the dual-purpose to diminish the gap by using very good bulls for milk, with a milk production index of about + 1000 kg (French system of evaluation). In that eventuality, the milk yield of the daughters would be increased by 500 kg. That is the reason why we analysed the situations where in L_1 and L_2, milk yield increased by 500 kg.

We supposed that the other performances (meat production, reproduction, milk composition) were unaffected which is certainly a rough hypothesis and an optimistic one since this calculation must be considered like a "last chance calculation" for the heavily dominated dual purpose breed.

In fact, it is well known that there is a genetic within breed antagonism between milk yield and :

- fat or protein % (international literature)
- reproduction (MOCQUOT and POUTOUS, 1975 on the French Friesian; TARTAR, 1980 on the very animals of Le Pin experiment).
- meat production : not very much for young bull production as has been demonstrated in the first results of the divergent selection in Le Pin, until the second generation (COLLEAU and LEFEBVRE, 1981) but much more for culled cows as demonstrated in Le Pin in the Norman population (COLLEAU, 1978b; COLLEAU and JOURNET, 1980).

In this option we, of course, took account of the increased food needs.

b. - <u>Basal food and system of distributing concentrate</u>

We will examine several situations for basal food allowing very different levels of milk production.

Food No 1 - basically constituted of maize silage of very good quality allowing the production of 16 kg 4% milk per day. This food is supposed to be distributed ad libitum if the daily production is above that level and to be adjusted to the needs if it is below (especially for the Norman cows that are more in risk of overfeeding with that food).

Food No 2 - Maize silage of less good quality allowing only 10 kg 4% milk.

Food No 3 - Hay allowing 4 kg 4% milk.

b.1 - In a first category of options, the curves of lactation were fixed at the curves observed in Le Pin, the concentrate compensating for the alimentary needs not satisfied by the basal food.

It can be verified that:

- The fact of having an improved Norman decreases the difference between the two breeds. It goes down from 800 to 380 kg equivalent milk per ha, on average for the three basal feed systems.
- The difference between Norman and Holstein does not depend very much upon the basal food quality, what is partly a consequence of our assumptions (feeding maize silage ad lib to Norman cows would give rise to problems in this breed and consequently lower the margin/ha comparatively to the Holstein breed). Nevertheless, from the results of the system No 3, it can be thought that the Norman cows would have an improved competitivity in the areas where good fodder, like maize, cannot be obtained (especially for climatic reasons).

b.2 - Further, it is likely that the system of concentrate allocation may also have an effect on the comparative profitability of the two breeds.

We "simulated" the effect of a 50% decrease in concentrate intake (with a refitting of the level of the milk yield permitted by the basal food since it is well known that there exists substitution phenomena between fodder and concentrate that vary according to the quality of the fodder). Such underfeeding leads to multiple consequences in fact : decrease of milk yield, decrease of fertility, increased interval between calvings, increased culling rate, decrease of beef value at culling and so on. To give the right figures to these phenomena is highly problematic and, any way, cannot be solved from our experimental results. So we carried all the consequences only on milk yield, by decreasing it according to the amount of the underfeeding and according to the conversion rates (Feed unit - Milk) given by the nutritionists. It is a rough approach and probably over-simple.

The results are not so surprising. With a basal level at 16 kg milk, the difference between Holstein and Norman is reduced but not by a large proportion (506 kg milk per ha vs 806 kg). With a basal level at 4 kg, the difference is greatly lowered: 125 kg vs 721 kg.

Finally, it can be seen that the difference of margin per ha between the two breeds may vary greatly in some of our "simulated" situations. The situations relatively favourable to the Norman breed are situations of low fodder quality and poorer management or low financial possibilities for the farmer. The simulated situations have to be experimentally verified and if they are reasonably true, we have to ask ourselves whether the condition favourable (or more exactly not very unfavourable) to the Norman are likely to be easily met in the future.

IV. - Influence of the Economic Environment

The previous results were obtained with certain price ratios representing a given economic environment. It is certain that changes in the price hierarchy may affect the differences of profitability between the two breeds.

To obtain trends, we only parametered the prices which had a strategic weight. As it was shown, the difference of margin was generally positive in favour of the Holstein breed and consequently we purposely only tested the conjectural hypotheses likely to narrow the gap, in order to investigate in what conditions the Norman could eliminate its disadvantage.

The major elements are milk price, beef price and concentrate price.

We tested the action of these elements in a good number of situations but in order to facilitate the presentation, we shall explain the results that would be obtained in our experimental conditions (basal level permitted without concentrate 7 kg milk/day) and with a combination of zootechnical parameters that seems not far from the present French dairy system, that is to say:

- Males and surplus females marketed as 8 day old calves.
- No beef crossbreeding.

- Culling rate : 25%.
- Calf mortality rate : 5%.

The results obtained are shown in Table 9.

a - <u>Influence of beef/milk price ratio</u>

The reference situation is characterised by a ratio:

$$\frac{\text{Price kg culled cows carcass weight}}{\text{price kg reference milk}}$$

which is equal to 12 or 10,5 for the Norman and for the Holstein, respectively.

We tested the incidence of a relative increase of beef prices of 10% comparative to milk prices; that increase was uniformily applied to each beef output : cows, male and female calves marketed for beef.

In this circumstance, the difference between the Holstein and the Norman decreases uniformly to 86 kg of equivalent milk per ha, for original Norman cows and for improved Norman cows. In terms of the relative decrease that result is especially interesting for the improved Norman cows.

b. - <u>Differential modification of beef prices for the two breeds</u>

The generalisation of Holstein animal and relative scarcity of Norman animals could lead to a relative improvement of the latter as far as beef prices are concerned.

That would affect every kind of cattle : calves and culled cows as well. However we simulated an improvement of Norman beef prices only for calf prices (+100FF). That seems reasonable and within the possible range of commercial reactions to the Holstein "tide".

Despite all these limitations, the effect is not negligible : it is comparable to a general 10% increase for beef prices (decrease of the margin difference by 65 to 90 kg).

c. - <u>Milk payment according to valuable matter</u>

By valuable matter, we mean the sum (fat yield + protein yield), each component of the sum being given the same economic weight.

Table 9 Influence of some parameters of the general economic situation on the difference of margin/ha (expressed in kg equivalent milk) between the two breeds

Situation (1)	Difference Holstein original Norman	Difference Holstein - improved (2) Norman
Reference	738	272
Beef prices : +10%	652	185
Price differences for 8 days calves increased by 100 FF	652	185
Milk payment according to valuable matter	264	- 24
Differential taxes on milk	566	143
Concentrate prices		
+ 10%	600	179
- 10%	875	364

(1) : Fixed zootechnical parameters

- no young bulls
- Surplus heifers marketed at 8 days
- no beef crossbreeding
- culling rate : 25%
- calf mortality rate 5%

(2) : Artificial Norman improved by 500 kg milk per lactation.

That system of payment is very different to the present French one. In some areas, different protein percentages are very little taken into account, if not ignored.

The result from Table 9 demonstrates that a major variation on the milk payment system is undoubtedly a major factor and could lead to very much improved living conditions for the dual-purpose breeds which have a high protein yield like the Norman breed. The difference of margin per ha would collapse (from 738 kg equivalent milk to 264).

d. - Differential taxes on milk

We tested the effect of political decisions intending to discourage a likely overproduction of milk. According to these, there would be a growing corresponsibility tax, depending on the overall amount of milk marketed by each farm. We took the following ratios : 0% up to 80.000 l, 1,5% between 80 and 150.000 l, 5% between 150 and 200.000 l, 10% beyond that point.

We worked on a farm area corresponding to 50 Holstein cows per herd, the surplus (males and females) being sold at 8 days. As expected, the milk price decrease is stronger in Holstein (2,5 c/kg) than in Norman (1,5 c/kg) and consequently, the margin difference per ha decreased by 170 kg equivalent milk, for the original Norman and by 130 kg equivalent milk for the improved Norman. The effect is thus stronger than the effect of improving beef prices, at least within our figures.

e. - Concentrate price

In our reference situation, the ratio

$$\frac{\text{price kg concentrate}}{\text{price kg reference milk}}$$

is equal to 1,25.

Market changes, especially on the protein fraction of the concentrate might lead to price variations. A +10% variation would have approximately the same effect as introducing the corresponsability taxes we imagined.

Conclusion

For many years, the dual-purpose breeds represented the selection objective in Western Europe, especially in France. That is now questioned, most of all since the Holstein's arrival. In France, not only is the Friesian breed increasing very rapidly but also the friesian is being more and more graded-up with Holstein genes.

It is difficult to contest that this breed may destabilise the national cattle production in the long run : too strong a movement toward the Holstein might lead either to milk surpluses which will be difficult to market, or to a dramatic decrease in the number of cattle and a corresponding decrease in beef production. A first calculation was made in France : the backward movement might be numbered in millions of cows (JARRIGE and LIENARD, 1976). The systematical development of beef crossbreeding then would have great difficulty in preventing by itself a drop of beef production, even if it were stimulated by every means.

In these circumstances, the dual-purpose breeds have still a role to play, provided that they are not being progressively given up by the dairy producer because they are less profitable.

The objective of this paper was to try to give an estimation of the profitability of the Holstein compared to the Norman breed, which is the major dual-purpose French breed, by means of experimental results coming from a research experiment mainly devoted to that purpose.

The presentation of the results is still rough, obviously, since the experimental results must acquire far greater accuracy and since methodology must be improved.

Our results confirm the difficulty for a breed like the Norman to narrow a milk gap of something like 1700 kg per lactation by means of its better milk composition, beef production and reproduction. The major items for competing better with the Holstein are the following:

- milk payment according to its actual industrial processing value, giving special weight to protein yield : giving up the present system of payment for a system based on valuable matter would be equivalent to improving the milk production of the Norman by about 500 kg per lactation in the present system, this would then reduce considerably the economic gap. Fortunately, that is the situation we are moving to in France.

- selection for milk production in dual-purpose breeds is an important point that must be stressed, not so much for decreasing the milk difference (impossible because Holstein is continuing to be improved and because the corresponding international active population is far larger) but for maintaining the gap at its present level. Such a result would be a success and it supposes that no possibility of selection pressure is lost towards by-side characters, except only of course for maintaining the absolute genetic level for beef production at its present level (that would be not too difficult).

Further, the dual purpose breeds are probably not so bad in relatively severe conditions (climatic and soil difficulties for growing very good fooders, high concentrate price...) for cows, our "simulated" results are questionable and we move towards a new experiment to study these problems. A clear unfitness of the Holstein breed in some circumstances has to be experimentally demonstrated, if it really exists. A clear answer would be helpful for guiding cattle policy.

References

Chevaldonne, M., 1978. La Holstein dans la zone Centre-Nord et l'Aube. in "La Vache Laitière" Ed.

Ed. INRA. Route de Saint Cyr - Versailles, 39-46.

Colleau, J.J., 1974. Comparisons entre la race mixte Normande, les races spécialisées Holstein canadienne et charolaise et leurs croisements. I. Performances de croissance des males et des femelles.

Ann. Génét Sél. anim., 6, (4), 445-462.

Colleau, J.J., 1975. Comparisons entre la race mixte Normande, les races spécialisées Holstein canadienne et leurs croisements. II. Performances d'engraissement et de carcasse des mâles.

Ann. Genet. Sél. anim., 7, (1), 35-48.

Colleau, J.J., 1978a. "Holsteinisation" de la population pie-noire française : prévisions de la fréquence des gènes Holstein dans les 20 années a venir.

Bull. Techn. CRZV - Theix - INRA (34) 45-50.

Colleau, J.J., 1978b. Comparison de types génétiques : bovins laitiers mixtes ou spécialisés. In "La Vache Laitière" Ed. INRA. Route de Saint Cyr - Versailles, 17-29.

Colleau, J.J., 1978c. Analyse et décomposition de la variabilité intergénotypique de la consommation alimentaire pour des jeunes bovins.

Ann. Génét. Sél. anim., 10 (1), 29-45.

Colleau, J.J., Journet, M., Hoden, A., Muller, A., 1979.

Feed efficiency during early lactation in cows of specialised and dual-purpose genotypes.

30th Animal Meeting of the E.A.A.P. - Harrogate, England.

Colleau, J.J., Lienard, G., Journet, M., 1979.

Comparison économique entre race Holstein et race Normande (Premiers résultats).

Bull. Techn. CRZV - Theix - INRA (35), 67-84.

Colleau, J.J., Journet, M., 1980. Relations entre variation de poids des vaches et production laitière.

Bull. Techn. CRZV - Theix - INRA (42), 25-30.

Colleau, J.J., Geay, Y., Malterre, C., Muller, A., 1980.

Effets d'une restriction alimentaire sur les performances d'engraissement et la composition des carcasses de taurillons Holstein.

Bull. Tech. CRZV - Theix - INRA (40), 13-17.

Colleau, J.J., Lefebvre, J., 1981. Divergent selection on milk yield in the Norman breed. I. Preliminary results on the indirect responses expressed by young bulls.

To be published in Ann. Génét. Sél. anim.

Dubis, J.M., 1979. Quelles vaches laitières pour demain? Comparaison de différents types.

Etude No 79063 de l'Institut Technique de l'Elevage Bovin - Paris.

Guittet, P., 1975. Comparaison entre Holstein et Frisonnes.

Mémoire de fin d'études E.S.A. - Purpan - p.65.

Jarrige, R. Lienard, G. 1976. Evolution du troupeau laitier et ses conséquences sur la production de viande bovine et l'avenir du troupeau de vaches nourrices.

Bull. Tech. CRZV - Theix, INRA (23), 47-59.

Journet, M., Hoden, A., Geay, Y., Lienard, G., 1973. Comparaison entre animaux Pie-Noire de type Holstein canadien et de type frison français.

CRZV - Theix, INRA (12), 13-26.

Poutous, M., Mocquot, J.C., 1975. Etudes sur la production laitière des bovins. III - Relation entre critères de production, durée de lactation et intervalle entre le ler et le 2ème vêlage.

Ann. Génét. Sél. anim., 1975, 7 (2).

Regis, 1979. Comparaison de l'aptitude à la production de veaux de boucherie de différents types génétiques pie-noirs.

Etude ITEB - GIE - Pays de Loire.

Regis, 1980. Comparaison de l'aptitude à la production de taurillons de différents types génétiques pie-noirs.

Etude No 80 034 de l'Institut Technique de l'Elevage Bovin - Paris.

Tartar, M., 1980. Relation mobilisation des réserves corporelles en début de lactation. Fertilité chez la vache laitière.

Mémoire fin d'études ENSA - Rennes, pp. 99.

DISCUSSION ON DR. J.J.COLLEAU'S & DR. G. LIENARD'S PAPER

Chairman: Prof. A. Neimann-Sorensen

Foulley: (a) The conclusion seems to be the opposite to the direction of your results, which indicate that the Holstein is better than the Normande.

(b) Do you suggest that the Normande should be used as a suckler breed, or as a dairy breed.

Colleau: (a) I do not conclude that the Normande is better, but rather I am talking about the survival of the Normande breed.

(b) A dairy breed, not a suckler breed.

Bittante: I note that both your trials are designed to measure the cows over a fixed productive lifetime (3 lactations in the first trial and 5 in the second). In Italy we have the largest population of Holstein and Canadian Friesians in Europe, where the majority of our Friesian cows belong to these two strains; we have found a reduction in productive lifetime and an increase in culling rate in comparison with our dual-purpose breeds. I think it is important in a comparative study to consider this characteristic; have you any data in France regarding this point ?

Colleau : The importation of Holstein genes is too recent in France to have accurate figures on the longevity of pure Holsteins as compared to European Friesians or dual-purpose breeds. Of course a first approximation can be made by using a reproduction criterion like the interval between calvings, which is worse in Holsteins than in Friesians or dual-purpose breeds. However, the most critical would, in my opinion, be the average number of lactations completed at culling. Using initial results on the first culled Holstein cows could lead to biased results, as pure Holstein cows in France are still rare, and therefore the breeders are likely to try to keep them longer at this point in time, than they would in the long-run.

Bittante: The use of maize silage, which is a very convenient foodstuff, has, in our environment, some disadvantages, often leading to reduction of average productive lifetime and to an increase in culling rate. This point should, in my opinion, be considered in future trials.

Colleau: In the trial I spoke about, the basal food was not a silage alone, but a mixture of maize silage, grass silage and hay, with pasture during the Summer. All these feeds were transformed using feed tables to maize silage equivalents.

Bech Andersen: Do you think you can find production systems severe enough to allow the Normande to become superior to the Holstein-Friesian ?

Colleau: Such a set of circumstances has not yet been found, but opinion is that the Holstein is not favoured by harsher conditions.

CHAIRMAN'S COMMENTS AT CLOSE OF SESSION III

Chairman : Prof. A. Neimann-Sorensen

In looking at strategies for future cattle breeding programmes we mostly have distinguished between two separate alternatives:

1) Either to try to combine - in an optimum way - milk and meat in the same animal - the dual purpose approach, or
2) to develop separate specialised breeds for dairy and beef production - the commercial crossbreeding approach.

There are pros and cons to both alternatives. In fact we should not think of the two possibilities as real alternatives. Both can be applied as a supplement to the other. There is not just one single strategy, it may depend on local circumstances.

In this session we have learned about the dual purpose approach, and seen that it is being practised and continuously being developed further in most countries. We have seen how such programmes can improve at the same time both the milk and beef potential of our dairy breeds.

One issue brought up this morning was the distinction between the breeding goals as determined on the farmer's level and on the national level. It was pointed out that the breeding objective should be determined on the farmer's level. This, I believe, is debatable, as the farmer's objecgives are always short-term, while national objectives are more long-term by nature. Also, cattle breeding itself is long-term; decisions made to-day will not be effective until 10 years ahead.

Finally, I want to stress the importance of having 'stable objectives' in our breeding work. By 'stable objectives I mean those which we can expect to have general and lasting value. Examples in my opinion would be such goals as feed utilisation, appetite for forages, etc. Such objectives mean better use of feed resources, thus improving the competitive power of the cow as a producer of food for humans.

USE OF BEEF BREEDS FOR CROSSING ON DAIRY HERDS IN THE NETHERLANDS

D. Oostendorp

Research and Advisory Institute for Cattle Husbandry,
Runderweg 6, 8219 PK Lelystad, The Netherlands.

Summary

The use of beef breeds for crossbreeding in the Netherlands is still limited. In 1980 the inseminations with semen of bulls of specialised beef breeds comprised only 1.1% of the total number of inseminations. Of the total of 21,000 beef inseminations, 44% were done with semen of Piedmont bulls and 29% with semen of double muscled bulls of the local breeds. The percentage of inseminations with semen of Charolais bulls has decreased to 5% because of the high incidence of calving difficulties with bulls of this breed.

Double muscled bulls are used by a particular category of farmer who is attracted by the very high prices of the calves and who takes into account the common use of caesarian sections. A further increase in the use of semen of double muscled bulls is not expected.

The majority of the dairy farmers are only interested in crossbreeding with beef breeds on condition that there will be no extra risks at calving. In this regard, results obtained with Piedmont bulls in recent years have been promising and there is a growing interest in the use of Piedmont bulls. Results obtained with Piedmont crossbred calves in bull beef and veal calf production systems have been good. However, the experience is based on a relatively small number of Piedmont bulls. Future developments will depend very much on the availability of an intermediate type of Piedmont bull, which combines a good beef production ability with guaranteed easy calving.

Introduction

In the Netherlands there has not been, and there still is not much interest in, the use of beef bulls on the Dutch dairy herd. The dairy herd of Dutch Friesian (DF), Meuse-Rhine-Yssel (MRY), and Groningen White head (GB) is well muscled and the calves are very suitable for veal production. The bull calves of the MRY breed are very acceptable for bull beef production. However, there is growing concern that the DF and MRY cattle are being influenced for the worse from the beef point of view by the introduction of North American Holsteins. A lot of data (Oldenbroek, 1981) indicate that the use of Holstein Friesian will have a big negative influence on the suitability of dairy cows and their offspring for beef production. At the moment attempts are being made to minimise this negative influence of the HF-breed by buying HF-bulls with a better fleshiness than the average HF-bull and paying attention to fleshiness in the performance tests of bulls. At the same time there is a growing interest in dairy farms for commercial cross breeding with beef breeds. But only under the condition that there will be no extra risk at calving.

Crossbreeding experiments

The research work on cross breeding began in the sixties with a 4-year trial carried out by Bergstrom (1973). Crossbred calves of Charolais, Limousin and Dutch Red and White sires and Dutch Friesian dams were fattened and compared with pure bred Dutch Friesian and pure bred Dutch Red and White animals. Semen was imported from 9 Charolais and 9 Limousin bulls. These bulls were selected in France for low incidence of calving difficulty. The sires of the Dutch breeds were chosen on their fleishiness. No heifers were included in the experiment. Male as well as female calves were fattened as veal calves, intensively fed young bulls and heifers and extensively fed steers and heifers. With regard to growth rates, feed conversion and carcass quality the Charolais crosses gave the best results whereas the crosses of the Limousin and the MRY were intermediate between the Charolais crosses and the pure Dutch Friesian. Although the Charolais bulls also gave the highest percentage of difficult calvings the results were judged acceptable for practical applications (Table 1).

On the basis of these results an insemination programme with Charolais semen was developed in practice. The calving results of these bulls have been reported by Van Leeuwen (1971).

During the practical application the number of difficult calvings rose to a level that was not acceptable to the average Dutch dairy farmer. This gave an enormous repercussion, not

Table 1 Percentage of difficult calvings and perinatal deaths in the cross breeding trial of Bergstrom (1964-1968).

	n	Difficult calvings incl. caesareans	Perinatal deaths
Ch x DF	166	9,2	3,0
Lm x DF	146	6,2	1,4
MRY x DF	137	5,1	3,0
DF x DF	-	1,8	3,7

Table 2 Percentage of difficult calvings and perinatal deaths in practice (Van Leeuwen, 1969-1970).

	n	Difficult calvings incl. caesareans	Perinatal deaths
Ch x DF	3000	18,1	4,7
DF x DF	-	3,7	2,2

only to the Charolais breed but generally to the possibilities of the beef crosses in the Netherlands.

In 1973/1974 another comparison between different beef breeds was made by Brunnekreef. In this trial semen was used from Fleckvieh, Chianina, Limousin, Belgian Blue and Piedmont bulls. Piedmont crosses came out very well with regard to calving difficulties (Table 3) and beef and veal production results (Brunnekreef 1975).

Introduction of the Piedmont breed

As a consequence since then a lot of attention has been paid to the possibilities of the Piedmont breed. The Piedmont, which originated from and is still largely confined to an area in north west Italy, is a medium sized rather than a heavy breed. It is claimed by the Italians that its doubled muscled character does not develop until well after birth and it is believed to be less likely to cause difficult calving than many other Continental breeds. For his trials

Table 3 Calving results in cross breeding trial of Brunnekreef (1973-1974)

Dam breed						
Sire breed	Limousin	Piedmont	Belgian Blue	Fleckvieh	Piedmont	Belgian Blue
Number of sires	4	4	3	4	4	3
Number of births	225	170	48	63	125	23
Gestation period (days)	286	287	280	284	285	281
Birth weight (kg)	42.2	46.3	48.6	43.6	43.3	46.9
Dead at birth (%)	2.2	2.1	2.0	3.1	0.8	4.3
Caesarean section (%)	-	1.1	4.2	1.6	3.2	-
Dead within 24 hours	3.0	1.1	4.0	1.6	1.5	-

Brunnekreef imported semen from 4 bulls recommended by the Experimental Station at Turin as proven not to cause calving difficulties when used on either Piedmont or Italian Friesian cows.

Because of the increasing demand for Piedmont semen, the AI-station in Heythuysen decided in 1976 to import their own bull Utente. Since then two other Piedmont bulls have been imported by this AI-station from Italy. Calving results following the use of semen of Utente were also good. The first check was done by Engels (1977). Out of 443 calvings of MRY-and DF-cows 2,7% caesarians and 3,7% perinatal deaths were registered. More recently calving results following the use of semen of different beef bulls by the AI-station West-Brabant were checked by De Koning (1980). In this case the Piedmont semen was from the bulls Utente and Barolo (also used in the Brunnekreef trial). From Table 4 it is clear that also in this case the Piedmont bulls came cut relatively favourable.

Table 4 Calving results AI-station West-Brabant (%) (De Koning 1980)

Sire breed	Pi	Li	Ch	DM-MRY
No of births	283	152	172	131
Very easy births	53,7	41,7	11,5	8,7
Normal births	38,2	45,5	48,3	37,0
Heavy births (1)	4,6	11,5	32,2	13,4
Caesarean births (2)	3,5	1,3	8,0	40,9
Difficult Births (1)+(2)	8,1	12,8	40,2	53,3
Dead within 24 hours	2,5	5,1	4,0	5,5

The suitability of the crosses of Utente for bull beef was checked by the Research and Advisory Institute for Cattle Husbandry on the Experimental Farm "C.R. Waiboerhoeve" and "De Vlierd" (Oostendorp, 1980). The results are summarised in Table 5.

Table 5 De Vlierd and Waiboerhoeve Young bull trials (Oostendorp 1978-1980)

Dam breed	MRY		DF	
Sire breed	MRY	Piedmont	DF	Piedmont
No of animals	34	52	40	23
Daily l.weight gain (kg)	0,97	0,93	0,93	0,89
Feed/l. weight gain	6,5	7,0	6,3	7,0
Carcass weight (kg)	319	355	285	332
Killing out%	57,7	60,8	57,7	60,6
IVO carcass classification				
- fleshiness	4	5-	4-	4+
- fat cover	3	3-	3	3-
- internal fat	3	3-	3+	3-
Price per kg carcass (D.fl)	7,35	7,80	7,15	7,50

The cattle were sent for slaughter when judged to be fit. The Piedmont crosses could be fattened to a higher slaughter weight without being too fat. The daily live weight gain did not differ very much. However, the killing out percentage for the crosses was greater by about 3% units and clearly the carcasses were more fleshy. The improved price per kg was partly due to this and partly to the fact that the carcasses were heavier. On the basis of the different experiments the difference in value of calves of beef cross breds was calculated in comparison with pure DF or MRY breeds. In Table 6 is shown that there is a big advantage in the use of the Piedmont crosses for bull beef production.

Table 6 Difference in value of calves of beef cross breds in comparison with pure DF or MRY breeds (D.fl.)

Trials	IVO	Hendrix Voeders	Res. and Adv. Inst. for C.H.	
			Waiboerhoeve	De Vlierd
HF x HF	-100			
HF x DF	- 50			
Pi x DF		+275	+245	+330
Pi x MRY		+180	+235	+350

Of course this advantage has to be divided between the dairy farmer and the beef producer on the basis of supply and demand. At this moment the price of Piedmont crosses (male and female) is about fl 100 above the price of the pure breds. On the calf markets as well as on the beef market a new market for this quality of animals has to be established.

Growing interest for Piedmont bulls in practice

Although there has been an increase in the use of semen from beef breeds the use is still of minor importance and formed in 1980 about 1,1% of the total number of inseminations.

However, from Table 7 it is clear that since 1975 the total number of inseminations of beef breeds has doubled, mainly due to the strong increase of use of semen of Piedmont bulls. In

Table 7 First inseminations in the Netherlands (Source: AI Year reports)

Bull breed	1975	1979	1980
Double muscled	4,567	7,751	6,349
Limousin	2,685	1,346	3,920
Charolais	2,628	1,972	970
Piedmont	516	5,762	9,479
Chianina	101	11	316
Belgian Blue	308	632	379
Fleckvieh (Simmental)	25	45	46
Total beef breeds	10,830	17,419	21,459
Total dairy breeds	1,617,891	1,816,671	1,875,616
Beef as % of total	0.7	0.9	1.1
NA Holstein as % total dairy	1.2	8.1	10.7

the same period the use of Charolais bulls has strongly decreased decreased. There was a slight increase in the use of semen of Limousin and double muscled bulls.

Most of the double muscled bulls are of the MRY-breed. Double muscled bulls are used by a particular category of farmers which is attracted by the very high prices of the calves. In this case the use of caesarian sections at birth is normally accepted. A further increase in the use of double muscled bulls is not expected (Bergstrom and Oostendorp, 1980).

On the other hand there is certainly room for a further increase in the use of beef breeds for cross breeding on the average dairy farm. In this regard the results of the Piedmont bulls have been promising. However, these results have been based on only a small number of Piedmont bulls that were not of the extreme double muscled type. The question is whether we will be able to select also in future sufficient bulls of the intermediate type that combine a good beef production ability with easy calving. If we succeed in this, more interest will certainly grow for this type of beef crosses on the average Dutch dairy farm. The condidition for a further progress are:

1. guaranteed easy calvings
2. guaranteed colour marking
3. guaranteed higher prices

At this moment the Piedmont corsses meet more or less these requirements.

References

Bergstrom, P.L. Gebruikskruising voor vleesproduktie bij rundvee.
I.V.O. rapport B 117, 1973.

Bergström, P.L. and Oostendorp, D. Muscular hypertrophy and its significance for beef production in the Netherlands.
EEC-Seminar Muscle hypertrophy in cattle, June 1980 Toulouse.

Brunnekreef, W. Uitkomsten geboorteregistratie en meskalverproeven van het Hendrix' proefkruisingsprogramma.
Bedrijfsontwikkeling 6 (1975) 402-406.

Engels, J. Onderzoek naar het geboorteverloop van de afstammelingen van de vleesstier Utente.
Rapport C.R.A. Roermond, 1977.

Koning, C.J.A.M. de. Redenen en resultaten van het gebruik van vleesrassen.
Rapport K.I. West-Brabant 1980.

Leeuwen, J.F.M. van. Resultaten van de geboorteregistratie betreffende de Charolais kruisings kalveren.
Bedrijfsontwikkeling 2 (1971) 49-50.

Oldenbroek, J.K. Meat production of Holstein Friesians in comparison to Dutch Red and Whites.
EEC-Seminar Beef production from different dairy breeds and dairy beef crosses, April 1981, Dublin.

Oostendorp, D. Developments in beef production in the Netherlands.
Irish Grassland and Animal Production Association Journal, 1980.

DISCUSSION ON IR. D. OOSTENDORP'S PAPER

Chairman: Prof. E.P. Cunningham

Diskin: Could you please comment on the inconsistency between percentage perinatal deaths and percentage difficult calvings in Table 1.

Oostendorp: I have no explanation, except that perhaps the sample size is small.

Fewson: Are the data in Tables 1 to 4 for heifers or older cows ?

Oostendorp: In all the experiments referred to, heifers have been excluded.

Smidt: Difficult births seem to be a major problem in dairy-beef crosses. I wonder, therefore, if there are any experiments going on in this context, either to bring the cows to term before they are due, by means of prostaglandins and/or corticosteroids, or to ease parturition by tokoylitic medication, e.g. by administration of β-mimetic drugs.

Oostendorp: Not to my knowledge.

Sreenan: I feel that the systems for induction of parturition being developed in New Zealand can be adapted for use in the context to which Professor Smidt refers. This system is based on the use of a combination of long and short acting corticosteroids administered at about 1 to 2 weeks prior to expected term date. We have used this combination to calve Charolais, Blonde d'Aquitaine and Limousin calves out of Hereford cross heifers following embryo transplantation. Calf viability was high and placental retention very low.

Thiessen: What were the breeds of the double-muscled group ? Did the Piedmontese and Belgian Blue breed groups also include double-muscled sires ?

Oostendorp: The double-muscled animals used were mainly MRY, but most breeds carried the double-muscled gene. We looked for double-muscled sires which were not extreme type.

Langholz: Does the quoted difference in calf prices between crossbred Piedmontese and purebred Friesian also stand for the female calves ?

Oostendorp: Yes, further trials have been conducted with heifers and a similar difference was also found for female calves.

Boucque: In Table 5 you cited a lower feed efficiency for the Piedmontese crossbreds, compared to the pure Dutch Friesians and MRY animals. Can this be explained by differences in the weight interval, that is where the Piedmont bulls are taken to a higher carcass weight. Normally, at equal slaughter weights we would expect the reverse, because the carcases of the crossbreds are leaner that those of the purebreds.

Oostendorp: Yes, the higher carcass weight is the major factor involved.

THE CURRENT PRACTICE OF COMMERCIAL CROSS-BREEDING IN THE UK WITH PARTICULAR REFERENCE TO THE EFFECTS OF BREED CHOICE

J. R. Southgate

Meat and Livestock Commission, Queensway House, Bletchley, Milton Keynes, MK2 2EF, United Kingdom.

Summary

It is estimated that around 30% of UK dairy cows are mated with a beef bull. The resulting crossbred cattle make a significant contribution to UK beef production both directly in dairy beef production and indirectly through the females making up a large proportion of the UK beef herd.

In dairy beef systems continental breed crosses increase total output but take considerably longer to reach slaughter condition. British breed crosses reduce both total output and the time taken to slaughter. Differences in feed efficiency are small. The time taken to slaughter is an important consideration in the UK where beef systems are matched to grass growth.

The medium size British beef breed cross Friesian cow is preferred to the continental cross cow in the UK beef herd, having a higher output of calf weight related to cow weight. Breed of bull effects in the beef herd are similar to those in dairy beef. Whilst breed differences in feed efficiency in the slaughter animal are small, overall breed differences in efficiency, involving the feed requirements of the suckler cow, favour the continental breed cross.

The structure of the United Kingdom cattle industry and sources of home produced beef

In 1978 approximately 1M tonnes of beef was produced in the UK, of this 58% was from animals bred in the dairy herd, 34% from the suckler herd and 8% from cattle imported from Ireland (Fig. 1).

The proportion of beef derived from cattle imported from Ireland is decreasing, due mainly to increased beef production there. In addition small changes in the size of the beef herd (Table 1) and an increased number of exported calves (Table 2) have contributed to a small decline in beef production (MLC,1981).

FIGURE 1 Structure of the UK cattle industry and sources of beef, 1978

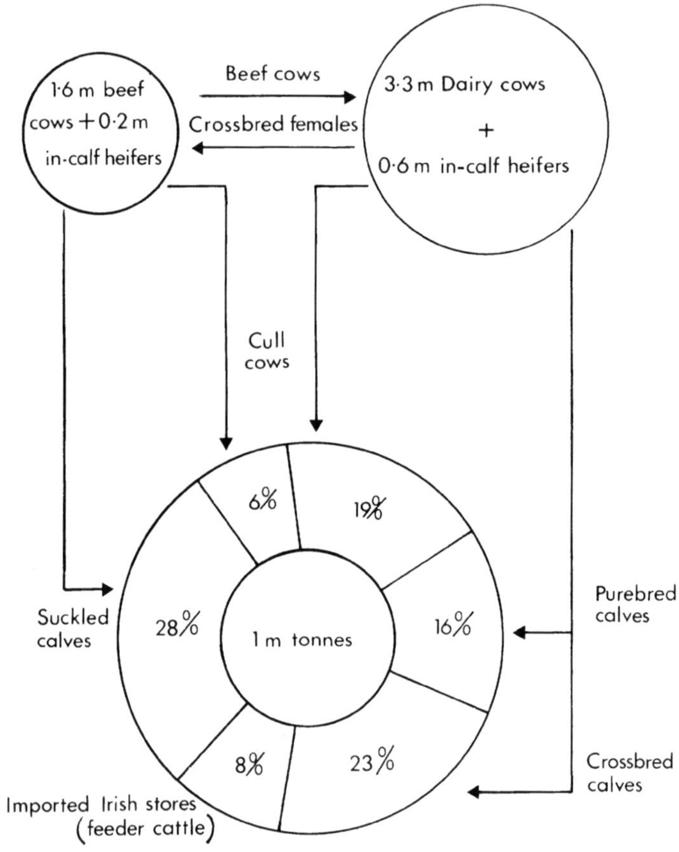

Source: MLC

Table 1 Numbers of breeding females in the United Kingdom, June

	1975 (000)	1976	1977	1978	1979	1980
Dairy herd cows	3242	3228	3265	3274	3292	3228
In calf heifers	664	713	634	678	684	677
Beef herd cows	1899	1764	1681	1588	1543	1479
In calf heifers	239	227	189	180	180	161

Source: MLC, 1981

The reduction in beef cow numbers will probably reduce the proportion of homebred beef from this source from 37% in 1978 to around 35% in 1980.

Table 2 Numbers of young calves exported from U.K.

	1975 (000)	1976	1977	1978	1979	1980
Heifer calves	13.9	10.5	16.2	16.4	17.3	21.4
Bull calves	116.4	239.3	378.6	404.3	347.7	277.2

Source: MLC, 1981

The Beef Breeding Herd

The UK beef herd is mainly crossbred. Although no census data are available on breed structure it is estimated that 36% of beef cows are by beef bulls out of Friesian cows; a high proportion of these by Hereford bulls. In general the cross between the small British breeds of bull and the dairy cow is popular as a suckler cow. Small size and good milking potential result in high calf weights per unit of cow size. (Table 3).

Hereford and Angus bulls predominated in beef herds until continental breeds became available. For many years the use of continental breeds was restricted by high bull prices. There was an increase in both the number and proportion of

Table 3 Weights of suckler cows and relationships with calf performance

Cow breed type	Cow weight* (kg)	Calf 200-day weight (kg)	Kg calf 200-day weight per 50 kg (kg)
Angus x beef breed	453	190	21
Angus x Friesian	449	193	21
Blue Grey	450	194	22
Charolais x beef breed	628	239	19
Hereford x beef breed	485	194	20
Hereford x Friesian	472	203	22
Red breed crosses	552	204	18
Shorthorn crosses	443	191	22

* Average of Spring and Autumn weighings

Source: MLC, 1979

inseminations undertaken in beef cows between 1964 and 1974, from 2.0 to 4.7% of inseminations (MMB, 1965; 1975), largely due to an increase in Charolais inseminations. However, the effect of this on the overall uptake of this breed in suckler herds should not be over-estimated. Legislative changes allowing the use of crossbred bulls and lower bull prices have been more important. It is estimated that in 1980 possibly 40% of calves from the beef herd were sired by Charolais bulls and a further 10% by other continental breeds (Kilkenny, 1980).

The change to Charolais bulls in suckler herds has resulted in an increase in calf sale weights (Table 5). While calf mortality and calving interval has also increased, the net effect of the change is an increase in herd output.

The U.K. Dairy Herd

Cow breed

The breed structure of the UK dairy herd has changed considerably since 1955 (Craven and Kilkenny, 1976 and Milk Marketing Board, 1963-1980). In England and Wales the proportion of

Table 5 Effect of sire breed on the productivity of Hereford x Friesian (H X F) and Blue Grey (BG) suckler cows

Sire breed	Charolais		Hereford		Angus	
Cow breed	HxF	BG	HxF	BG	HxF	BG
Calf weaning weight* (kg)	294	278	249	234	236	220
Live calves per 100 cows calved	94.9	95.6	98.2	98.7	98.5	98.9
Calving interval for cows that calved (days)	378	369	372	367	370	366
Calf weight per cow put to bull (kg/year)	269	265	240	223	231	217
Calf weight per 50 kg cow weight (kg/year)	26.9	29.3	23.9	25.0	23.1	24.0

* Calf weight adjusted to 250 days

Source: MLC, 1979

Shorthorn cows has fallen from 25% in that year to less than 1% in 1979. The proportion of the small-sized breeds (Ayrshire and Channel Island) has fallen from 36% of the cow population to 8%, whilst the Friesian has increased from 41% in 1955 to 89% in 1979. The trend is similar in Scotland and Northern Ireland, at the expense of the Ayrshire and Dairy Shorthorn, respectively.

Whilst the main reason for this change was undoubtedly the milking ability of the Friesian, there can be no doubt that the superior beefing quality of the Friesian, particularly when crossed to beef breeds, augmented this change and is an underlying reason for the current interdependence of the beef and milk industries.

In the past few years the move towards greater total milk yield per cow has continued with an increased use of North American Holsteins. In 1979 pure Holsteins accounted for only 1% of the dairy cow population (Milk Marketing Board, 1979). However, it is estimated that about 20% of the 'black and white' calf crop now contain Holstein blood. It is likely that this trend will continue.

Incidence of cross-breeding

Crossbreeding, as a means of upgrading to a new breed became established practice in dairy breeding. Although the overall change in the breed structure of the dairy herd has been dramatic it has been achieved gradually and accomplished using only a porportion of the breeding herd. There has been, quite naturally in the prevailing climate, a tendency to crossbreed cows not required for breeding dairy replacements.

Survey data (Milk Marketing Board, 1963-1980) has shown that while there has been a five/six year cycle in the extent to which diary cows have been crossbred to beef bulls, there has been no overall trend from 32% (Fig. 2).

Beef breed use in crossbreeding

Trends in beef breed use in the dairy herd since the early 1960's have not been the same as in the suckler herd. The Hereford has maintained a steady popularity, accounting for around 60% of beef inseminations. Initially, the Charolais made more rapid advances than in the beef herd - made possible by the greater use of AI in dairy herds - but has levelled out at around 16% of beef inseminations. In the beef herd, both Angus and Hereford use has declined as Charolais increased. However, in dairy herds, the increase in use of the Charolais has been associated with a decline only in the use of Angus. This has not been a simple substitution. In the years when the proportion of Charolais, and later as Simmental and Limousin inseminations have increased, Hereford inseminations have decreased but Angus usage has been hardly affected. These periodic increases in the use of the continental breeds have always been associated with periods when confidence in dairying is increasing and the proportion of cows inseminated with dairy breeds is rising. Thus in this situation the Hereford has accounted for a lower proportion of a reduced demand for beef inseminations though it still dominates.

In practice, Angus semen is used mainly on heifers, while continental breeds are used only on mature cows. In a generally declining demand situation, the Angus has only maintained its level of usage in years when dairy insemination are greatest. This probably reflects a greater proportion of maiden dairy heifers in the sample in these periods. The increased use of continental breeds has occurred mainly in the mature cow population and has been directly responsible for a decline in Hereford usage. However, Hereford insemination levels have always recovered when the trend has swung back to beef.

The Simmental made fairly rapid gains in the early 1970's. In part this may have been due to the extensive publicity given to the breed in these years, however, to a large extent it has been replaced by the Limousin - which created a demand much more slowly. This slower establishment of the Limousin may well have been caused by a lower level of interest in experimenting with 'new' breeds after the initial reaction to the UK Government's less restrictive policy towards importation of the early 1970's wore off and the mid 70's slump in world trade took effect.

Limitation to beef crossbreeding in dairy herds

Cunningham (1974) has calculated that there are practical limits to crossbreeding to beef bulls in dairy herds where the size of the herd is to be maintained. This limit is a function of the herd life of cows. At the average level of crossbreeding to beef bulls in the UK (32%), it can be estimated that dairy herd size can be maintained providing cows average more than 3.5 lactation. In practice, many small herds need to breed a higher number of replacements than are theoretically necessary. Thus the level of crossbreeding to beef bulls in UK dairy herds is probably near the practical maximum. Changes in the size of the dairy herd are achieved practically by altering the herd life of cows as well as by breeding policy. In effect, therefore, the ability to increase herd size is not limited by the level of crossbreeding to beef in the short term.

Effects of crossbreeding on UK beef production

Benefits to corssbreeding are due to heterosis and the combining of attributes of more than one breed in a single animal.

Heterosis effects in beef production are well documented. They are generally beneficial, particularly in traits concerned with reproduction. However, the use of crossbreeding in the UK dairy herd is probably close to its optimum level and within the commercial beef herd it is almost universal. It would appear therefore, there is little further benefit to be gained from heterosis effects within the UK related to the F_1 hybrid.

Within the beef herd there is still some use of backcrossing - mainly involving the Hereford - which does not realise the full potential of heterosis effects.

An analysis of calf 200-day weights in suckler herds (Table 6) showed that where a Hereford bull was used, the superiority of the calves out of Hereford X Friesian cows over those out of Blue Grey cows was less than when other breeds of bull were used.

It is difficult to estimate the extent to which this backcrossing occurs; but the increased use of the Charolais in the beef herd is probably reducing it.

Breed effects in the beef herd

Cow breeds

Calf performance in terms of 200-day weight is affected by cow breed (Table 3). Reliable comparisons are restricted to different crossbred types as purebred cows are too few in commercial beef herds. The larger breed types produce the heavier calves at 200 days. Cows with Friesian blood produce heavy calves - not only because of the large body size of the Friesian but also because of the effect of its superior milking potential on calf growth.

Calving interval is related to cow size (Table 7) with the large Charolais x Friesian having an interval eight days longer than the small Angus x beef cow.

Dystokia and perinatal calf mortality have only been

Table 6 Contemporary comparison of 200-day weights of suckled calves out of Hereford x Friesian (HXF) and other beef cows by different sire breeds

Sire breeds	+ Differential (kg)	Probability
Lowland herds:		
*Larger than Hereford	+24 ± 4	0.001
Hereford	+18 ± 3	0.001
**Smaller than Hereford	+24 ± 5	0.001
Upland herds:		
ᵀLarger than Hereford	+18 ± 4	0.001
Hereford	+12 ± 3	0.001
Smaller than Hereford	+20 ± 3	0.001

+Difference (±) of HxF over beef cows

*Includes Charolais, South Devon, Lincoln Red, Sussex and Devon.

**Includes, Angus, Shorthorn, Galloway and Welsh Black.

Source: MLC, 1979.

Table 7 Effect of cow type on calving interval

Breed type of cow	Mean calving interval (days)
Angus crosses	372
Angus x Friesian	374
Blue Grey	367
Hereford x Friesian	376
Charolais crosses	378
Charolais x Friesian	380
Hereford crosses	372

Source: MLC, 1979

reliably recorded in herds containing Hereford x Friesian and Blue Grey cows (Table 8). Both assisted calvings and calf deaths were greater in the larger Hereford x Friesian cow by 1.2 and 0.8% units, respectively.

Table 8 Assisted calvings and calf deaths in suckler herds (cow matings)

	Breed of cow			
	Hereford x Friesian		Blue Grey	
Breed of sire	% assisted calvings	% calf deaths	% assisted calvings	% calf deaths
Charolais	10.1	5.1	9.0	4.4
Simmental	9.7	4.7	8.9	3.7
South Devon	8.4	4.4	6.3	3.7
Limousin	7.9	4.9	6.4	3.9
Lincoln Red	6.0	3.5	4.8	2.4
Devon	6.1	3.2	5.0	2.0
Sussex	4.0	1.7	2.7	1.3
Hereford	4.2	1.8	3.3	1.3
Aberdeen Angus	2.0	1.5	1.1	1.1

Source: MLC, 1979

Whilst the increased calving interval and higher calf mortality associated with larger cow types reduces annual output in terms of calf weight, the growth rate contributed to the calf by the larger size of cow still results in a net positive relationship between cows size and annual output of calf weight (Table 3). In terms of energetic efficiency, however, the larger cow may not be optimum. In UK conditions, the calf output per unit of cow weight is maximised with the medium weight cow (Table 3).

Sire breeds

The output of beef herds is affected by sire breed (Table 9).

the larger breeds producing the heaviest calves at 200 days.

Table 9 Effects of sire breed on calf 200-day weights

	Lowland	Upland	Hill	Weight above (+) or below (-) Hereford x calves in same herd (kg)
	Weight at 200 days (kg)			
Charolais	240	227	205	+15
Simmental	232	222	198	+13
South Devon	231	221	200	+12
Devon	225	215	191	+ 9
Lincoln Red	222	214	189	+ 8
Limousin	215	204	186	+ 8
Sussex	215	207	186	+ 5
Hereford	208	194	184	0
Angus	194	182	176	-14

Source: MLC, 1979

However, as with cow breed type, bull breed size is also associated with higher levels of dystokia, assisted calvings, calf mortality (Table 8) and a longer calving interval (Table 10).

The net effect is still a positive relationship between bull breed size and annual herd output (Table 3).

Breed effects in suckled calf finishing systems have been investigated in the MLC Breed Evaluation Programme (Southgate, Kempster & Cook, 1980 and Kempster, Southgate and Cook, 1981). In this programme autumn-born suckled calves were purchased at one year of age and finished during the winter on a complete diet. The cattle were fed individually and slaughtered when estimated subcutaneous fatness (SF_e) was 8%, using an ultrasonic scanner to aid selection. Similarly, spring-born suckled calves were purchased at seven months, overwintered at around 0.5 kg daily gain and then finished in the summer in the same way as the autumn-born calves.

Table 10 Effect of breed of sire on calving interval

Sire breed	Days to next calving	Days from previous calving
Angus	370	371
Charolais	375	373
Hereford	371	372
South Devon	376	374
Lincoln Red	374	372
Sussex	372	371
Devon	374	372

One side from each animal was subjected to a detailed cutting test involving the preparation of retail joints, deboned and trimmed to a standard fatness specification (saleable meat).

Breed difference in calf output from the suckler herd are also reflected in differences in slaughter weight (Table 11 & 12) to the extent that Charolais calves were slaughtered 27% heavier than Angus calves at the same SF_e. The large difference in age at slaughter accounted for 66% of the difference in slaughter weight between Charolais and Angus crosses. These breeds were also of very different weight at the start of the finishing period and this accounted for 28% of the difference in slaughter weight. Even though the Charolais crosses had a daily gain in the finishing period 10% greater than Angus crosses this had a very small effect on slaughter weight and only accounted for 6% of the difference.

Feed intake in the finishing period varied markedly between sire breeds, but differences in feed efficiency (in terms of liveweight gain) were small and not significant.

Calves out of Hereford x Friesian cows, which were 12 kg heavier at the start of the finishing systems than those out of Blue Grey cows, were also heavier at slaughter. Daily gains were similar as were ages at slaughter.

Table 11 Least square means for performance traits

Winter finishing of autumn-born suckled calves out of Hereford x Friesian and Blue Grey

Sire breeds	No.	Weight at start (kg)	Weight at slaughter (kg)	Age at slaughter (day)	Total feed (kg)	Daily feed intake (kg)	Daily gain (kg)	Feed efficiency (g gain/kg feed)
Angus	60	319 b	393 e	477 c	927	8.9 c	0.77 bcd	86 a
Charolais	61	363 a	494 a	520 a	1520	10.2 a	0.84 abc	82 a
Devon	61	328 b	419 cd	494 d	1094	8.9 c	0.78 abcd	88 a
Hereford	59	322 b	410 d	492 d	1066	8.8 c	0.78 abcd	88 a
Limousin	48	332 b	454 b	517 a	1348	9.2 bc	0.78 abcd	85 a
Lincoln Red	60	336 b	428 c	489 d	1121	9.6 b	0.85 ab	89 a
Simmental	62	359 a	490 a	517 a	1485	10.2 a	0.86 a	84 a
South Devon	59	335 b	451 b	517 a	1342	9.2 bc	0.77 abcd	84 a
Sussex	59	324 b	428 c	512 a	1255	8.9 c	0.76 cd	85 a
Cow breeds								
Hereford x Friesian	320	340 a	445 a	504 a	1246	9.4 a	0.80 a	86 a
Blue Grey	286	328 b	429 b	502 a	1196	9.2 b	0.78 a	85 a

Sire breeds means with the same suffix did not differe significantly at the 5% level of probability.

Source: MLC Breed Evaluation Programme.

Table 12 Least square means for performance traits
Summer finishing of spring-born suckled calves out of Hereford x Friesian and Blue Grey cows

Sire breed	No.	Weight at start (kg)	Weight at slaughter (kg)	Age at slaughter (kg)	Total feed (kg)	Daily feed intake (kg)	Daily gain (kg)	Feed efficiency (g gain/kg feed)
Angus	72	317 d	408 e	505 e	967	10.1 c	0.96 b	95 a
Charolais	70	355 a	525 a	570 a	1775	11.1 ab	1.07 a	96 a
Devon	70	339 bcd	444 d	518 d	1086	9.9 c	1.00 ab	100 a
Hereford	72	328 cd	442 d	527 d	1176	10.0 c	1.00 ab	100 a
Lincoln Red	71	352 ab	465 c	520 d	1196	10.9 b	1.05 a	97 a
Simmental	71	352 ab	526 a	572 a	1860	11.4 a	1.08 a	95 a
South Devon	72	342 abc	488 b	551 b	1513	10.8 b	1.06 a	98 a
Sussex	72	334 c	461 c	538 c	1306	10.2 c	1.03 ab	101 a
<u>Cow Breeds</u>								
Hereford x Friesian	327	346 a	478 a	538 a	1232	10.7 a	1.04 a	97 a
Blue Grey	243	334 b	461 b	537 a	1186	10.4 b	1.03 a	98 a

Sire breeds means with the same suffix did not differ significantly at the 5% level of probability.

Source: MLC Breed Evaluation Programme.

Table 13 Least squares means for carcass traits

Winter and summer finishing of autumn and spring born suckled calves respectively out of Hereford x Friesian and Blue Grey cows

			Killing out %		Saleable meat in carcass %		% saleable meat in higher priced cuts	
Sire Breed	Winter	Summer	Winter	Summer	Winter	Summer	Winter	Summer
Angus	60	72	52.5 b	51.0 c	72.5 bc	72.0 a	44.1 c	44.0 bc
Charolais	61	70	54.8 a	52.7 a	72.7 b	71.9 a	44.8 b	44.8 a
Devon	61	70	52.7 b	51.0 c	71.6 d	71.3 bc	44.0 c	43.9 c
Hereford	59	72	52.3 b	51.1 c	71.9 c	71.9 cd	44.1 c	44.4 b
Limousin	48	-	54.7 a	-	73.3 a	-	45.4 a	-
Lincoln Red	60	71	52.3 b	50.9 c	70.8 e	70.7 c	44.3 c	44.1 bc
Murray Grey	15	-	53.4	-	72.0	-	44.3	-
Simmental	62	71	53.0 b	52.2 ab	72.0 cd	71.7 ab	44.8 b	44.7 a
South Devon	59	72	53.2 b	51.8 b	72.0 cd	71.2 bc	44.3 c	44.1 bc
Sussex	59	72	53.1 b	51.8 b	72.6 b	71.6 ab	43.9 c	44.1 bc
Cow breed								
Hereford x Friesian	320	327	53.4	51.7	72.2	71.4	44.4	44.2
Blue Grey	286	243	53.0	51.5	72.1	71.5	44.4	44.3

Sire breeds means with the same suffix did not differ significantly at the 5% level of probability.

Source: MLC Breed Evaluation Programme.

Breed effects in dairy beef production

Cow breed effects

The predominance of the Friesian breed during the 1960's and 1970's has diminished the commercial relevance of cow breed effects. Purebred Ayrshire and Channel Island cattle perform so badly in beef systems that most of those not required for breeding are slaughtered as young calves. To some extent, crossbreeding these breeds to Charolais has increased their acceptability, but they are still inferior to the cross out of the Friesian (MLC, 1979).

To a large extent, the present structure of the UK beef industry, involving the interdependence of dairy and beef production is due to the beefing quality of the Friesian. Southgate (1981a) has shown that the Holstein is of inferior quality for beef production and the increasing level of inclusion of Holsteins in the UK dairy herd may adversely affect this structure. The present subjective standards applied at all levels of the beef industry discriminate heavily against the Holstein or Holstein type and, if continued, may lead to a larger proportion of the calves out of Friesian type cows being rejected for beef production and slaughtered.

Sire breed effects

Dairy beef systems have also been included in the MLC Breed Evaluation Programme. Growth and feed intakes have been recorded for the pure Friesian and crosses by the important beef breeds out of the Friesian in production systems lasting 16 and 24 months. The 16-month system, in which growth rate is essentially linear, approximates to the 18-month yard finishing system which is commercially popular in the UK. In the 24 month system yearlings are subjected to a store period lasting nine months during which growth rate is restricted by feeding a low quality feed. This system is similar to the commercial two year grass finishing system in the UK. The carcass in these trials were subjected to the same cutting tests as those from the suckled calves.

Sire breed effects in dairy beef systems closely resemble those in suckler beef (Table 14 & 15). Cattle sired by the continental breeds were superior for carcass quality and yield in terms of killing-out percentage, saleable meat percentage and the distribution of saleable meats in the higher-priced cuts. Of the British breeds, Angus and Sussex crosses had similar saleable meat percentage to the Charolais and Simmental crosses.

In terms of growth rate and slaughter weight the breed differences recorded in these trials are very similar to those recorded on commercial farms by MLC, (1979).

Implications of breed choice

Suckler beef - output of saleable meat

Breed differences in suckled calf production and finishing systems are large and the number of suckler cows required to produce a given amount of saleable meat annually is a function of the following parameters:

(i) Percentage of saleable meat in carcass
(ii) Killing-out percentage
(iii) Liveweight at slaughter
(iv) Calf survival
(v) Calving interval

In the UK suckled calf production is in two distinct phases - calf production and finishing. Whilst an overall evaluation of breeds is important, an evaluation within each of these sectors is required. The weight of calves at the point of transfer from the production to the finishing sector is an important practical parameter.

By using estimates of these parameters from data presented in this paper it can be shown that more Angus cross cattle are required to yield a given amount of saleable meat than Charolais crosses. Calving interval and calf survival are adversely affected by the larger breeds. None-the-less fewer cows are required for mating to a Charolais than to an Angus to give a similar annual yield of saleable meat (Table 17). It is not clear what effect environment has on this. MLC

Table 14 Least squares means for performance traits - 16 month beef

	No.	Start weight (kg)	Slaughter weight (kg)	Slaughter age (days)	Daily gain 3 mths to slaughter (kg)	Feed consumption +(kg) 3 mt slaughter	Feed efficiency+ -(g gain per kg feed 3 mt slaughter
Friesian (F)	47	85 b	438 d	501 b	0.85 b	3244	108
Angus x F	24	78 c	371 f	439 c	0.84 b	2564	115
Charolais x F	32	99 a	510 a	524 a	0.94 a	3733	108
Devon x F	57	84 b	387 e	441 c	0.86 b	2567	120
Hereford x F	56	89 b	400 e	455 c	0.85 b	2575	120
Simmental x F	33	96 a	489 b	507 ab	0.93 a	3498	110
South Devon x F	24	95 a	457 c	492 b	0.90 a	3276	111
Sussex x F	29	87 b	406 e	461 c	0.86 b	2738	118

Sire breed means with the same suffix did not differ significantly at the 5% level of probability.

+ Only breed group feed intakes were recorded to 11 months and consequently multiple range test were not applied to overall feed data.

Source: MLC Breed Evaluation Programme.

Table 15 Least squares means for performance traits 24 month beef

	No.	Start weight (kg)	Slaughter weight (kg)	Slaughter age (days)	Daily gain 3 mth to slaughter (kg)	Feed consumption +(kg) 3 mth slaughter	Feed efficiency+ (kg) (g gain per kg feed 3 mth to slaughter)
Friesian	46	91 bc	542 c	738 a	0.70 bc	6212	72
Angus x F	22	87 bc	454 e	685 b	0.63 d	5153	72
Charolais x F	30	105 a	616 a	743 a	0.79 a	6418	78
Devon x F	61	89 bc	496 d	705 b	0.67 c	5358	76
Hereford x F	43	86 bc	495 d	698 b	0.68 c	5182	79
Simmental x F	32	98 b	569 b	745 a	0.72 b	6363	73
South Devon x F	23	94 bc	554 bc	733 a	0.72 b	6043	77

Sire breed means with same suffix did not differ significantly at the 5% level of probability.

+ Only breed group feed intakes were recorded to 20 mths and consequently multiple range tests were not applied to overall feed data.

Source: MLC Breed Evaluation Programme.

Table 16 Least squares means for carcass traits - 16 and 24 month dairy beef

	Number		Killing out %		Saleable meat in carcass (%)		Percentage saleable meat in-higher priced cuts	
	16	24	16	24	16	24	16	24
Friesian (F)	47	46	49.6 bcd	49.2 cd	70.2 c	69.7 c	44.5 ab	43.8 a
Angus x F	24	22	48.5 d	48.7 d	71.4 ab	71.3 a	44.5 ab	44.0 a
Charolais x F	32	30	51.5 a	51.6 a	71.6 a	70.8 ab	44.9 ab	44.0 a
Devon x F	57	61	48.5 d	49.0 d	70.7 bc	70.2 bc	44.6 ab	43.8 a
Hereford x F	47	43	49.1 cd	50.1 bc	70.8 bc	70.1 bc	44.4 ab	43.9 a
Simmental x F	33	32	50.4 b	50.4 b	71.4 ab	70.3 bc	45.1 a	44.3 a
South Devon x F	24	23	50.1 b	50.7 b	71.4 ab	70.5 bc	44.2 b	43.3 b
*Sussex x F	29		49.7 bcd		71.9 a		44.2 b	

Breed means with the same suffix did not differ significantly at the 5% level of probability.
* Insufficient on Sussex x F cattle in 24 month beef.

Source: MLC Breed Evaluation Programme.

data shows (MLC, 1979) that the breed effects on 200-day weights are smaller in hill than lowland herds; the difference between Hereford x Friesian and Blue Grey cows being reduced from 21 kg to 9 kg. If calf mortality and calving interval respond in a different manner to calf weight, there may be an interaction of breed and environment on annual output.

Table 17 Estimate of the number of suckler cows required to provide one tonne of saleable meat per year

Sire breed	Angus		Charolais	
Cow breed	Blue Grey	Hereford x Friesian	Blue Grey	Hereford x Friesian
Cows required	6.9	6.8	5.4	5.4
Calves required for finishing	6.8	6.5	5.1	5.0
Saleable meat yield (kg)	1000	1000	1000	1000

(Estimates from Tables 7,8, 10 & 13).

There are only small breed differences in the distribution of saleable meat between the higher and lower priced cuts (Tables 13 & 16). Within a system, the largest difference was between Angus and Limousin crosses in winter finishing. Even where saleable meat in the high priced cuts is worth twice that in the lower priced cuts, the difference between these breeds amounts to less than 1% of the total saleable meat value.

Suckled calf feed requirements

In the UK the preference for medium-sized crossbred cows - either out of dairy breeds or beef breeds - diminishes the practical implications of cow breed differences for feed intake. The weight difference between the popular breeds is quite small (Table 3) and in the absence of precise feed intake data, cow weight is probably the best guide to feed intake. In addition to the cows requirements, the calf has a steadily increasing feed requirement up to the time it is transferred to the finishing unit. Again, precise data are not available and average calf weight is probably a reasonable guide to calf feed intake. Whilst it is recognised that there is an interaction between cow feed intake, milk yield and calf feed intake, it seems reasonable to expect the feed intake effects to be inversely related so that cow weight plus average calf weight can be used as an indicator of feed requirement with some confidence. In Table 18 cow/calf feed requirements have been estimated for Angus and Charolais bulls used on Hereford x Friesian and Blue Grey cows. Feed requirements have been expressed in hectares, using average stocking rates for autumn calving herds (MLC, 1980) adjusted for average concentrate usage. This method estimates that cows mated to Charolais bulls require only 4% more feed than those mated to Angus bulls. However, as Angus cross calves yield less saleable meat and more cows are therefore required, the feed requirement of cows mated to Angus bulls is 18% greater on an equal saleable meat yield basis.

Precise estimates of feed requirements for finishing suckled calves are available from the Breed Evaluation Programme (Tables 11 & 12). Breed differences in feed efficiency are small (Tables 11 & 12) and in situations where these data are not available, close estimates of feed requirements can be made as simple functions of daily gain and the length of the finishing period. Table 18 shows the total area required to finish Angus and Charolais crosses out of Hereford x Friesian and Blue Grey cows estimated from the total feed intake results in Table 11 apportioned one third to concentrates at 4t/ha and two thirds to forage at 11.0 t/ha. On this basis Charolais crosses require 64% more feed per head than Angus crosses. However, because of their greater yield per head of saleable meat, Charolais crosses require only 33% more feed in finishing systems than Angus crosses on an equal yield basis.

Charolais crosses require 10% less feed than Angus crosses when the feed requirements of both the calf production and finishing sectors are combined.

Table 18 Total feed requirement of suckler cows and calves to slaughter to provide one tonne of saleable meat

	Angus		Charolais	
Cow breed	Blue Grey	Hereford x Friesian	Blue Grey	Hereford x Friesian
Cow/calf requirement to *transfer per cow (ha)	0.67	0.69	0.69	0.72
Cows required (Table 17)	6.9	6.8	5.4	5.4
Total cow/calf requirement * to transfer (ha)	4.6	4.7	3.7	3.9
Calf requirement per head - finishing (ha)	0.14	0.14	0.23	0.23
Calves required (Table 17)	6.8	6.5	5.1	5.0
Total finishing requirement (ha)	0.9	0.9	1.2	1.2
Total system requirement (ha)	5.5	5.6	4.9	5.1

* transfer from calf production to finishing unit.

The superiority of the Charolais cross in overall feed efficiency is fairly robust to changes in the relationship of stocking rates in the calf production and finishing sectors. Where the stocking rate in the finishing system assumed in Table 18 is doubled, the feed requirement of the Charolais cross is only 5% less than the Angus cross; where the stocking rate is halved, the feed requirement is 14% lower.

The reduction in total system feed requirement with the Charolais cross is due to the much reduced cow/calf feed requirement to calf sale (18%) offset by the 25% greater feed requirement in the finishing part of the system. This large difference in feed requirement in the finishing system is mainly a function of the time taken to reach slaughter point.

At the point where the Angus crosses were slaughtered, Charolais crosses had consumed only 17% of the overall difference in feed. Similarly, the large difference in total output of the two breed types is largely a function of time to slaughter. At the time the Angus crosses were slaughtered the Charolais crosses, whilst being 51 kg, heavier, had only gained 7kg more in the finishing period.

In the example, the cow breeds used have only a minor effort on productivity and feed use. Whilst this may reflect the UK situation other crossbreeding strategies may include large differences in cow/calf feed requirements and give greater importance to cow breed effects.

Similar exercises can be undertaken for different suckler beef production systems. Whilst the use of resources will be functions of a similar set of parameters, it must be appreciated that it is assumed that feed requirement and availability are the same. This may be more or less true nationally, but in the individual farm situation there is a need to match systems to resources. In particular, grazing systems have to balance grass requirement to grass growth. Care must be taken that the different time requirements of breeds to reach an acceptable slaughter point are not beyond the limits of the grazing season.

It would not seem from the example that there is any conflict in the choice of breed within the calf production and finishing sectors. However, there is no doubt that the crossbreeding strategy which tends to minimize cow feed requirements whilst maximising calf growth is a definite objective for the calf producer. The finisher on the other hand must relegate breed choice to a lower order of importance. He must be more aware of his resources and appreciate how different breeds complement them. Maximising output per animal is not so important because of the small differences between breeds in the basic efficiency of converting feed to meat.

Dairy beef

In the UK it is reasonable to allocate all cow costs to the

dairy industry. Thus the use of resources in dairy beef is a simple function of those parameters affecting suckled calf finishing. Values for these parameters are given in Tables 14, 15 and 16.

Dairy beef systems are an aggregate of successive, seasonally based, rearing periods. In all but the ultimate, finishing period growth rate and efficiency are the main determinants of output and feed requirements. However, in the finishing period, breed differences in time taken to slaughter have by far the largest effect on output and feed requirement. In winter finishing systems these breed differences can be practically accommodated by altering the balance between grazing, conservation and concentrate feeds. However, in the grass finishing system, which relies on grass growth, the larger, later finishing breeds will not be ready for slaughter before grass growth declines and will need either supplementary feeding at higher cost or a further yarded period. Whilst this can be overcome by using a calf born earlier in practice this is aggravated by calf supply or price problems.

Provided breed choice does not exceed these practical limitations, the continental crosses give the highest output per head and per breeding cow. However, because of the lack of breed difference in feed efficiency, breed choice by dairy beef producers is more likely to be influenced by compatibility with seasonal and market factors. The dairy producer, on the other hand must offset the beef producers breed demand against breed effects within the dairy herd. The use of the larger continental breeds involve more dystokia which is probably associated with loss of milk production, longer calving interval and higher mortality. These disadvantages have effectively limited the use of continental breeds in dairy herds and have resulted in price differentials for young calves which effectively reduce breed choice by the beef producer.

References

Craven, J.A. and Kilkenny, J.B., 1976. The structure of the British cattle industry.
In Principles of Cattle Production. Edited by H. Swan and W.H. Broster, Butterworths, London. 1-44.

Cunningham, E.P. 1974. The economic consequences of beef crossing in dual-purpose or dairy cattle populations.
Livest. Prod. Sci., 1: 133-139.

Kempster, A.J., Southgate, J.R. and Cook, G.L., 1981. The carcass characteristics of suckler-bred cattle of different breeds and crosses.
Anim. Prod. 32:2: (Abstract - in print).

Kilkenny, J.B., 1980. Personal communication.

Meat and Livestock Commission, 1979. Data Summaries on beef production and breeding. Rvsd Dec 1979.
MLC, PO Box 44, Bletchley, Bucks, UK.

Meat and Livestock Commission, 1981. UK Market Review 1980/81.
MLC, PO Box 44, Bletchley, Bucks, UK (in print).

Milk Marketing Board, 1963 to 1980. Reports of the breeding and production organisation. MMB, Thames Ditton, Surrey, UK. 17 to 20.

Milk Marketing Board, 1979. Dairy facts and figures 1979. The Federation of United Kingdom Milk Marketing Boards, Thames Ditton, Surrey, UK. 36.

Southgate, J.R., Kempster, A.J. and Cook, G.L., 1981. The growth performance of suckler-bred cattle of different breeds and crosses.
Anim. Prod. 32:2: (Abstract - in print).

Southgate, J.R., 1981a. A note of the comparison of Holstein and Friesians for growth, feed efficiency and carcass traits.
EEC Seminar, Dublin 1981. EEC Brussels.

DISCUSSION ON MR. J.R. SOUTHGATE'S PAPER

Chairman: Prof. E.P. Cunningham

Fiems: Calf mortality appears to be very similar between breeds in your results; other research workers have reported bigger differences. Is this an effect of age of cow in your study, or does anyone have any information on this subject.

Southgate: We have observed in other data (not presented in this paper) that there appears to be an optimum age in relation to calving difficulty; young and old cows have a higher incidence of calving problems when compared to cows in their middle lactations.

Cunningham: The general rule appears to be that first calvers have about two-and-a-half times more calving difficulties than mature cows. It is worth noting the complexity of the structure of the population in the United Kingdom, which this paper illustrated. This offers considerable scope for exploiting breed differences.

BEEF CROSSING ON THE DAIRY HERD - WHY DOES IT WORK IN THE BRITISH ISLES AND NOT IN GERMANY ?

H.J. Langholz

Institute of Animal Husbandry and Genetics,
University of Gottingen, Federal Republic of Germany.

Summary

Though net efficiency of commercial crossbreds sired by bulls of large-framed breeds is found to be distinctly improved under experimental field conditions and though severe attempts have been made at least by some AI studs to introduce commercial crossbreeding into Friesian breeding, this method does not come to a real breakthrough in the Federal Republic of Germany. Reasons for this situation may be deduced from a comparison of the German situation with that of the British Isles, where crossbreeding in dairy herds forms an essential part of cattle production strategy. The following facts seem to be of the utmost importance :-

- problems of within-herd maintaining of purebred heifer numbers due to small herd sizes and to temporarily higher demand for replacement;
- increased frequency of calving difficulties and pre- and post-natal losses after crossbreeding with large-framed sires, especially with those not thoroughly tested for this trait;
- less favourable economic prospects of specialised beef production with heifers including little chance of crossbred heifer calves to be utilised for replacement in suckler herds, as compared to raising purebred heifer calves for breeding purposes.

As long as these main obstacles are relevant to the German cattle industry, greater prospects for commercial crossbreeding are to be expected only among dual-purpose breeds whose female offspring are promising replacements for dairying in unfavourable environments. An increasing percentage of crossbreeding Friesian cows with sires of the Red and White breed (in Lower Saxony, e.g. the percentage is at present 4.3%) indicates that breeding under farm conditions already moves into this direction.

The situation

The economic evaluation of the comprehensive EEC funded crossbreeding trial in Germany reveals that gross margin increases by up to 25% in crossbred bulls sired by bulls of a large-framed breed provided that feeding intensity is sufficiently high and/or fattening is extended up to high final slaughter weights. This is, e.g., obtained by a 600-day indoor feeding regime on maize silage or by an 850-day grazing indoor finishing system. Both systems lead to final weights of over 600 kg.

Increase in gross margin in crossbred heifers is about 10%. This is valid for finishing from pasture after the second grazing period. This system turns out to be best both in terms of net profit and product quality.

Despite this evident superiority in fattening performance of commercial crossbreds over purebred Friesians and despite the severe attempts made at least by some A.I. studs to introduce commercial crossbreeding in Friesian breeding, this method does not really come forth in the Federal Republic of Germany.

As shown in Table 1, commercial cross inseminations with beef-type sires even in the north-west region of the FRG with more dairy-type cattle populations do not exceed 3% and significantly decreased during recent years.

Table 1 Inseminations by specialised beef sires and "others"

Region		1970	1972	1974	1976	1978	1979
North-West	SH, NS, NRW	9 157	21 781	45 632	35 242	32 731	30 360
		.7%	1.5%	2.9%	2.0%	1.7%	1.5%
Central	H, RHP, S	-	1 972	7 170	10 012	8 914	9 599
		-	.5%	1.5%	2.0%	1.7%	1.8%
South	BW, BY	1 576	4 674	2 092	2 169	1 615	1 452
		.1%	.2%	.1%	.1%	.1%	.1%
Total	FRG	11 333	28 427	54 894	47 423	43 260	41 411
		.3%	.7%	1.2%	1.0%	.8%	.8%

North-West : Schleswig-Holstein (SH); Niedersachsen (NS); Nordrhein-Westfalen (NRW).

Central : Hessen (H); Rheinland-Pfalz (RHP); Saarland (S)

South: Baden-Wurttemberg (BW); Bayern (BY).

As the German A.I. statistics do not differentiate between breed of sire and breed of dam, figures in Table 1 include all inseminations made with specialised beef breeds and "others" resulting in over rather than underestimated figures. However, not recorded in Table 1 are cross matings between Red-and-White and Friesian practised apparently to an increasing extent. Indirect calculations result in 4.3% cross matings between Friesian cows and Red-and-White sires in Lower Saxony in 1978/79 (Table 2).

Table 2 First inseminations in German Friesians of Lower Saxony by breed of sire in 1978/79

	R E G I O N		
Breed of sire	Weser-Ems %	Hanover %	Lower Saxony %
Friesian	94.7	94.6	94.6
Red-and-White	5.0	3.8	4.3
Others*	0.3	1.6	1.1

* Mainly specialised beef breeds (Charolais)

In Great Britain, however, crossbreeding for beef from dairy herds meanwhile forms an essential part of cattle production strategy. Percentage of dairy cows inseminated by beef bulls fluctuated between 30 and 40% and has always been somewhat higher in the more specialised Ayrshire breed than in the British Friesian breed. However, in recent years there has been a steady decline in crossbreeding Friesian cows, i.e. from 35% in 1974/75 to 25% in 1978/79. This seems to indicate long-term structural changes in the British cattle industry. Hereford and Angus sires have been for years the prevailing sire breeds used for crossbreeding in Great Britain. Recently, however, larger-framed sire breeds, especially Charolais and Limousin, gained significantly in point.

Factors affecting the implementation of commercial crossbreeding for beef

A number of reasons may be responsible for the different acceptance of crossbreeding in these two areas of the European Community. From our experience the following facts may be considered the main obstacles which make the North German dairy farmers reluctant to adopt crossbreeding procedures:

- Small dairy herd sizes together with the preference for self-replacement of their herds.
- Increased frequency of calving difficulties and calf mortality after crossbreeding with large-framed sire breeds.
- Less favourable results with female calves when used for beef production than when used for raising breeding heifers.

Furthermore, the following facts are likely to raise additional problems when innovating commercial crossbreeding into dairy cattle breeding:

- Insufficient sensitivity in calf marketing and inadequate marketing system for trading quality calves for the specialised beef industry.
- Low degree of specialisation in beef production, especially regarding beef production with suckler cows.
- Inadequate exploitation of higher beefing capacity, especially in North-West Germany due to insufficient roughage quality and feeding procedures.
- Prolonged length of gestation after crossbreeding and its negative effect on the milk yield per year.
- Strong tradition of purebreeding in the German cattle industry.

Effect of herd size

In the FRG the average size of dairy farm is small. This adversely affects the application of crossbreeding when replacement is practised on a within herd basis. According to the 1979 census, average cow number per dairy herd in Germany is 11.3. This is four times less than the average number of dairy cows in British dairy herds (Table 3). Model calculations by ZICKGRAF (1978) indicate that commercial crossbreeding in self-maintaining dairy herds applies to larger herds only (Table 4). Extending crossbreeding to 40% would decrease the probability of accomplishing the required replacement to 70% even in herds of 20-40 cows. On the other hand, only herd sizes of 100 cows and above would allow one to extend crossbreeding to about 1/3 without greater risks of falling short of own replacement.

Table 3 Cattle production structure in selected EEC member countries (1979/80)*

	FR Germany	France	United Kingdom	Irish Republic
Farm size ha	15.0	25.9	50.4	20.5
% cattle farms	64	62	66	83
% dairy farms	52	46	25	48
% dairy cows	97	73	69	77
cattle/100 ha	114	73	74	107
cattle/farm	28.8	33.9	84.9	28.6
dairy cows/farm	11.9	13.6	48.9	14.2

*derived from EEC Dairy Facts and Figures 1980. Economic Division MMB.

Table 4 Percentage of possible commercial crossbreeding in self-replacing herds in relation to the probability of self-replacing and herd size (ZICKGRAF, 1978)*

Probability of self-replacing %	Herd size n	No of calves required n	No of purebred matings required n	% of possible crossbred matings %
70	20	6	13	40
	40	11	24	40
	100	28	59	42
	350	96	198	43
90	20	6	16	20
	40	11	28	30
	100	28	65	36
	350	96	210	40
99	20	6	22	-
	40	11	35	13
	100	28	75	26
	350	96	226	35

* 4 lactations/cow

Effect of increased calving problems and calf mortality

Increased feeding intensities in beef production and market preference for heavy lean carcasses favour the use of large-framed sire breeds for commercial crossbreeding in Germany. Without consistent selection for ease of calving, frequency of dystocia after crossbreeding with large-framed sires is 2-3 times the average of that after purebreeding, as observed in both series of our crossbreeding trials (Table 5). In the same order of magnitude also peri and postnatal calf losses increased, except in Red-and-White crosses. This phenomenon, however, might be due partly to the very unfavourable hygienic field conditions under which the German series of commercial crossbreeding trials were conducted. In the United Kingdom, crosses sired by Hereford and Angus bulls in the majority of cases do not raise such problems. Furthermore, the significant role of crossbreeding in A.I. services enables the implementation of comprehensive testing programmes. In Germany, bulls offered by the A.I. studs for crossbreeding purposes in general are not progeny-tested either for calving performance or for beef production capacity.

Prospects for the utilisation of female crossbreds

The third just as important obstacle to introducing commercial crossbreeding into dairy cattle breeding in the FRG is the limitation to the utilisation of female crossbreds in specialised systems of beef production.

Main obstacle is that crossbred calves cannot be used for replacement in suckler herds, since the percentage of beef produced by cow-calf operations is still very low in the FRG amounting to about 1.5% of total beef production. On the other hand, in Great Britain and also in Ireland an important part of the total cow population is kept as suckler cows (31 and 23% respectively in 1979, Table 3), the majority of which (95%; CUNNINGHAM, 1978) are crossbred cows. According to MLC field recording results, weaning performance of crossbred suckler cows is significantly better than that of purebred beef cows, increases being respectively 5.9, 5.2 and

Table 5 Reproductive performance after crossbreeding German Friesian cows with beef sires

Frequency of dystocia			Losses during first 10 days			Losses during rearing period		
	I	II		I	II		I	II
GF	1.6%		GF	2.4%		GF	6.8%	
GRW X	3.4%		GRW X	1.4%		GRW X	6.5%	
Ch X	4.4%	2.6%	GY X	1.4%		GS X	11.0%	
Chi X		2.8%	Ch X	4.2%	1.5%	Ch X	11.6%	3.9%
GS X	5.0		GS X	5.0%		GY X	17.7%	
GY X	5.1%		Ma X		2.6%	Chi X		10.0%
Ma X		3.8%	Ro X		2.6%	Ro X		11.8%
Ro X		7.6%	Chi X		4.3%	Ma X		17.0%

GF = German Friesian, GRW = German Red-and-White, Ch = Charolàis, Chi = Chianina
GS = German Simmental, GY = German Yellow, Ma = Marchigiana, Ro = Romagnola
I = 1st series
II = 2nd series.

1.6% for lowland, upland and hill conditions and showing interactions with sire breeds (MLC, 1971). Also under marginal grassland conditions in the Central German hills region, crossbred suckler cows from commercial crossbreeding between respectively Charolais and Simmental x Friesian achieve distinctly better reproductive performance than purebred breeds used for suckler beef production. Preliminary results from the ongoing EEC-funded project reveal heterotic effects of 15% for Simmental x Friesian crosses (Table 6).

On the other hand, it has to be realised that a high demand for purebred Friesian heifers improves the competitive situation of purebred matings. Drastic structural changes in cattle production together with an increasing concentration of dairy cows in the coastal grassland region and genetic changes towards more Holstein type Friesian cattle led to increased demand for heifers. Comprehensive exports of purebred heifers to North Africa and the Near East released additional demand for this category of cattle. In this connection the above mentioned decreasing percentage of crossbreeding in the British Friesian herd might be explained by current changes in the breeding policy towards higher-yielding more dairy-type cows, thus causing temporarily an increased demand for purebred heifers for herd replacement.

Additional factors restraining commercial crossbreeding

Among the additional factors restraining commercial crossbreeding in the FRG insufficient procedures of marketing crossbred calves must be mentioned first. There is no quality recording on the North German calf market as, e.g., in Great Britain by the separate quotation for crossbred calves. Trade tradition of 10 day old calves does not allow the introduction of centralised calf markets because of hygienic reasons which could be disadvantageous mainly for the transfer of crossbred calves sired by bulls of large framed breeds as experienced in our trials.

The less refined marketing procedures in the North German calf trade seems to be connected with the low degree of

Table 6 Calving percentage of single-crossbred suckler cows at the experimental station, Relliehausen

		Breed of cow			
		GF x GF	CH x GF	F1 x GF	F1 x F1
1st calving	n	60	56	49	43
	%	85.7	87.5	92.5	78.2
2nd calving	n	40	44	29	20
	%	80.0	93.6	78.4	62.5
3rd-6th calving	n	57	74	46	17
	%	78.1	85.1	93.9	77.3
Average of 1st-6th calving	n	157	174	124	80
	%	81.3	87.9	89.2	73.4

specialisation of cattle farms. 79% of all cattle farms in the FRG are still keeping also dairy cows, while in the United Kingdom this value amounts to 38% only (Table 3). Increasing specialisation of bull beef production based on maize silage in the long run is likely to react on the male calf market. Recent increases in price differences between breeds clearly indicate this situation (Figure 1). However, there are no signs of a developing market for female calves for beef production.

While in the United Kingdom the majority of crossbreds sired by Hereford or Angus bulls fit well for the traditional steer or heifer beef production system, commercial crossbreds sired by bulls of large-framed breeds in Germany can only be exploited according to their genetic potential in fattening systems providing them a sufficient feeding intensity and carrying them on to high final weights. Insufficient nutrient concentration in the roughage components, improper feeding management and seasonal limitations to feeding are the main reasons why crossbred bulls under average North German fattening conditions do not exhibit their maximum growth potential. On pasture, carcass gain of respectively Charolais and Simmental x Friesian crossbreds was only 5.5% superior, while under maize silage fattening conditions superiority grew to 9.5%.

Except in Red-and-White crosses, gestation length after crossbreeding with sires of large-framed breeds is prolonged by 4.2-7.6 days in the order Simmental<Romagnola<Chianina< Yellow cattle<Charolais<Marchigiana. The direct effect on gross margin per dairy cow is negligible, being 0.40 DM/day for a family farm and 13.0 DM/day for a farm working with hired labour (ZEDDIES, 1976). More severe losses might be caused by recovery problems following difficult calvings. However, no consistent analyses have been carried out so far to elucidate this situation.

Finally, crossbreeding of cattle has no tradition in Germany. This is in contrast to the situation in Great Britain and Ireland where crossbreeding on a commercial scale has always been an acknowledged tool of the cattle industry.

Future prospects of commercial crossbreeding in Germany

As long as suckler beef production systems play only a minor role in the German beef industry, which could absorb all the commercial female crossbreds for replacement, and as long as other specialised systems of fattening female calves remain less profitable, commercial crossbreeding with specialised beef breeds in the FRG will obviously not gain in point.

Therefore, attention should be focussed more on systems of crossbreeding between different types of dual-purpose breeds whose offspring could be used for milk production in less favoured production environments. Simmental x Friesian crossbred cows might be a competitive alternative for indoor or paddock production systems in the southern regions of the FRG, while Red-and-White x Friesian crossbred cows might better fit to the adverse grassland conditions of the coastal area of the central uplands of this country. The steady and remarkable increase in crossbreeding Friesian cows to Red-and-White bulls must, at least to some extent, be interpreted as a counter-reaction to the increasing specialisation for dairy traits in Friesian breeding.

Further, serious attempts should be made to improve marketing of high-quality calves for beef production. The introduction of central markets for trading 5-6 week old starter calves sold on a per kg liveweight price basis seems to be the most promising approach to serving the specialised beef sector. In general, an essential pre-requisite is the thorough testing of sires to be used for crossbreeding in A.I. services on a larger scale which in the case of limited demand for such sires per breeding unit should be organised in inter-regional or even international co-operating schemes rather than in own "mini-test programmes".

References

Diehl, R.F., 1976. Vergleichende Untersuchung zur Fortpflanzungsleistung nach Einfachkreuzung beim Rind. Diss. Göttingen.

Meyer, J.-N. 1969. Die Bedeutung von Umweltreizen fur das Brunstund Paarungsverhalten landwirtschaftlicher Nutztiere. Diplomarbeit Göttingen.

von Stietencron, D. 1978. Wirtschaftliche Erfolgsaussichten einer Rindfleischerzeugung mit Einfachkreuzung unter Praxisbedingungen des norddeutschen Raumes. Diss. Göttingen.
Zeddies, J. 1976. Der wirtschaftliche Wert der Fruchtbarkeit. 6. Hülsenberger Gespräche. Schriftenreihe der Schaumann-Stuftung zur Förderung der Agrarwissenschaften, p. 12-33.
Zickgraf, -. 1978. Wieviel Kuhe kann man innerhalb einer Herde für Gebrauchskreuzungen freigeben ? Der Tierzüchter 30, 46-48.
NN 1980. EEC Statistics on Agricultural Markets. Livestock Products. ISBN 92-825-1905-8.
NN 1981. Markt. DLG-Mitteilungen. 2nd issue, p. 93.
NN MMB Breeding and Production. Annual Statistics No. 26-29. Milk Marketing Board, Thames Ditton, Surrey, England.
NN EEC Dairy Facts and Figures. 1980. Economic Division, Milk Marketing Board, Thames Ditton, Surrey, England.
NN Rinderproduktion in der Bundesrepublik Deutschland. Annual Statistics. Arbeitsgemeinschaft Deutscher Rinderzüchter. Bonn 1970, 1972, 1974, 1976 and 1979.
MLC 1971. Beef Improvement. Scientific Study Group Report. Special Issue of MLC.

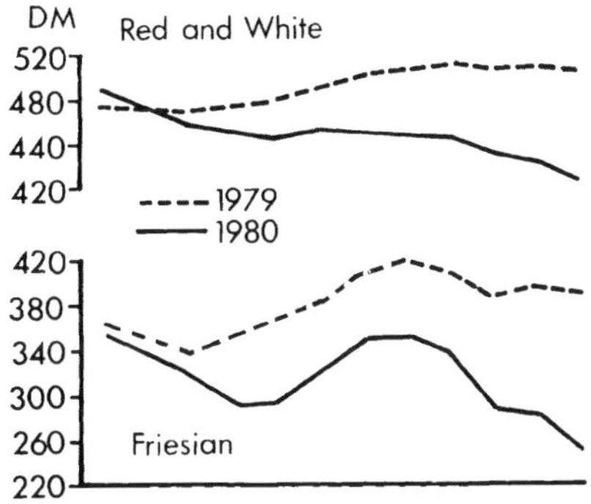

FIGURE 1 Market prices for male rearing calves North-West Germany (up to 14 days old)

DISCUSSION ON PROF. DR. H.J. LANGHOLZ'S PAPER

Chairman: Prof. E.P. Cunningham

Neimann-Sorensen: In your interesting analysis of the reasons why crossbreeding is not practised in West Germany (and other Continental countries) I think there is one factor which has not been sufficiently emphasised; that is, the importance of attaining high yields in the dairy herd. It is not incidental that we have a culling rate of 40% in many dairy herds; it is a matter of economic necessity. The dairy farmer feels that he cannot have less productive cows in the herd and he therefore wishes to bring in all heifers to allow maximum selection in the herd. A culling rate of 40% leaves no room for crossbreeding.

Langholz: In Germany, the main reasons for the lack of crossbreeding are (1) structural changes in our cattle set-up, giving a high demand for replacement heifers; (2) the rapid change to Holstein Friesian creating a high demand for heifers; (3) the demand for dairy heifers from the oil-producing countries. On the other hand, this situation should lead to the introduction of crossbreeding in dual-purpose breeds, which would have more stable milk yield, especially under poor conditions.

Liboriussen: The use of these dual-purpose beef breeds does allow a farmer the option of keeping on females for breeding stock. The genetic differences between animals within breeds are large enough to allow considerable selection within breeds.

Langholz: Yes, but it raises the question whether we wish to select within one breed or develop two strains. Also, there are heterotic effects to be gained from crossing. Furthermore, two-thirds of the gain from using a Charolais sire can be obtained on maize feeding using dual-purpose crosses (Red and White crosses).

Oostendorp: I would like to come back to the point made by Neimann-Sorensen on the factors influencing beef crossing rate. Another factor is the question of economics, for example, the price of milk. It is interesting to see that the percentage of beef crossing in Denmark dropped after their entry into the Common Market.

Neimann-Sorensen: We have a 40% culling rate and it would take a big change in milk-to-beef price ratio to change this.

Langholz: In Denmark, dairying is based on fodder crops rather than grass. In Germany, we have a big difference between the longevity of cows in grazing areas (4 lactations) and those regions using a lot of by-products (around 3 lactations).

Cunningham: In Ireland there are three main reasons why beef crossing is so successful: (1) the high prices paid for beef calves compared to Friesians - 50% more for Continental crosses and 20% more for Hereford crosses; (2) the production per cow is low and the average number of lactations is five, so provision of replacement heifers causes little difficulty. The seasonal nature of production - 85% of cows calve in the three months of spring. Furthermore, cows calve at two years of age and farmers are not interested in retaining replacements from late calvers; progeny tested dairy bulls are used early in the mating season, followed by beef bulls in the second half of the season.

CURRENT EFFICIENCY OF EMBRYO TRANSFER PROCEDURES AND THEIR POTENTIAL FOR INCREASING BEEF OUTPUT FROM THE DAIRY HERD

J. M. Sreenan, & M. G. Diskin

Agricultural Institute,
Belclare, Tuam, Co. Galway

Summary

Recent progress in the development of embryo transfer technology is outlined and possible ways in which this technology might be used to increase beef output from the dairy herd are presented.

The overall effect of increasing the reproductive rate by superovulating the top producing cows (top 10% of cows) combined with single embryo transfers would not increase the proportion of beef cross calves because of the necessity to use the remainder of the herd as recipient cows for the embryos.

The induction of twin-calving, by embryo transfer, in a portion of the dairy herd would of course increase beef output.

Various possible breeding strategies involving twinning in the dairy herd and the resulting breed composition of the calf crops are presented.

Introduction

Despite the fact that a very large number of potential fertilizable eggs (up to 75,000) are present in the calf ovaries from birth, yet the reproductive capacity of the cow is limited. This limitation is due to the fact that at the end of each oestrous cycle only one egg is normally matured and shed and following successful breeding and a gestation length of 280-290 days this will result in at best an annual production of only one calf. This combination of single ovulations and the long gestation length results in an average production of 6 calves per cow lifetime.

Embryo transfer procedures involving increases in ovulation rate followed by the harvesting, storing and transferring of the resulting embryos to recipient or host cows are aimed at effecting a significant increase in the reproductive rate of the cow. This increase would be effected in two ways; firstly by the ovulation rate increase in any donor cow and secondly by inducing a proportion of recipient cows to carry twin calves.

Current level of Technology

Embryo transfer as a total procedure involves the following steps:-

1. Superovulation or the hormonal stimulation of the donor cow ovaries to shed larger numbers of eggs at the end of an oestrous period.
2. Breeding of the donor cow to fertilize these eggs.
3. Recovery of the fertilized eggs or embryos from the donor cow uterus.
4. Storage of the embryos for short periods at 37°C or extended periods at -196°C.
5. Transfer of the embryos to recipient cows whose cycles are synchronous with that of the donor cow.

Superovulation

The currently used procedures for the most part involve gonadotrophin (FSH like compound) administration at mid-cycle (Days 10, 11, or 12) with prostaglandin administered 48 hours later to induce luteal regression, oestrus and ovulation.

The endogenous level of LH released at the beginning of the induced oestrus is usually adequate to cause ovulation of most of the extra stimulated follicles and there is no evidence to suggest that augmentation with exogenous LH has any advantage in the ovulation process. Because LH is, however, involved in normal follicular maturation a small proportion may be included when FSH is the gonadotrophin administered.

Superovulation rates from current experiments are shown in Table 1. The gonadotrophins used were pregnant mare's serum gonadotrophin (PMSG 2,000 i.u.), pregnant mare's endometrial cup secretion (PMEG 2,000 i.u.) and a combination of FSH/LH (ratio of 5:1); 30 mg FSH and 6 mg LH. Gonadotrophin was administered on Day 10 followed by PG on Day 12. Mean ovulation rates varied between 13-17 ovulations per donor and was higher following the pregnant mare's gonadotrophin.

However, a better distribution (5-20) of ovulations with a lesser overall variance was recorded following the FSH/LH combination. These ovulation rates are similar to those being currently recorded in the literature (Boland et al., 1981). Variation in individual donor response to superovulation treatment is still the greatest problem. In the study referred to in Table 1., response ranged from 0-53 ovulations.

Embryo Recovery

Non-surgical embryo recovery procedures are currently in use in most research and commercial units. Recovery is normally effected through the use of 3-way Foley type catheters inserted cervically. The major factor affecting non-surgical recovery rates is certainly operator skill and experience. An efficient operator will recover an average of 60% of the shed eggs or embryos.

Non-surgical recovery has the advantage of being a repeatable technique without causing any trauma to the donor cow.

Embryo Storage

For short-term storage periods (24 hrs or less) it is possible to keep embryos alive in a variety of buffered salt solution supplemented with serum (cow or sheep) or serum albumin. Such storage is normally at temperatures of 24-35°C.

For long-term storage, deep freezing procedures for cow embroys have been developed (Wilmut & Rowson, 1973, Willadsen et al. 1978) based on storage at -196°C in liquid nitrogen.

Survival rates of frozen-thawed embroys are shown in Table 2 from current studies at this laboratory. Survival here is based on continued development from the morula to expanded or hatched blastocyst in culture following thawing.

Over 80% of Day 7 embroys survived the freezing process, as determined by in vitro culture, and this is in agreement with the data of Willadsen et al.(1978).

However, the ultimate test of viability for the embryo is the pregnancy rate achieved following trnasfer. Pregnancy rate following the transfer of frozen-thawed embroys is given

Table 1 Superovulation Response

		PSMG (2000 i.u.)	PMEG (2000 i.u.)	FSH/LH (30mg/6mg)
No. of Donors		18	15	15
Mean No. Ovulations (±S.E.)		17.2±3.3	16.9±2.0	13.5±1.8
Fertilisation rate		85%	82%	83%
Ovulation distribution	0 - 4	22%	0%	7%
	5 - 20	33%	67%	73%
	21+	44%	33%	20%

Table 2 Embryo Freezing - Survival Rate following culture

	Embryo age (days) 6	7	8
No. embryos thawed	35	78	29
No. developed _in vitro_	0	61	11
% Survival	0%	78%	38%

Optimum freezing/thawing procedure (D_7)

Survival rate ---------------- 48/56 (86%)

Table 3 Embryo Transfer - Pregnancy Rates following various embryo storage and transfer methods

Flank 'Fresh' (1+CL)	Flank 'Frozen' (1+CL)	Non-surgical 'Fresh' (1+CL)	Non-surgical 'Fresh' (1-CL(+A.I.))
73%	61%	50%	60%
46%		36%	45%
87%		63%	40%
66%		60%	86%
75%		45%	
70%	61%	48%	62%

+CL = ipsilateral to Corpus Luteum

-CL = contralateral to Corpus Luteum

in Table 3. The pregnancy rates following the transfer of frozen-thawed embroys have been variable and the higher rates recorded may be due to greater selectivity of morphologically viable embroys at transfer.

The ability to successfully freeze embroys gives much greater flexibility. Recipient cows or heifers need not be assembled until the exact number of embroys available is known. This of course greatly reduces the costs involved.

Embryo transfer

Because embroys are normally collected at Day 6 or later, transfer must be to the uterus of the recipient heifer or cow. Transfer into the uterus may be effected in two ways, viz., via the Flank or non-surgically via the cervix.

Flank transfer

The uterus is exteriorised through an incision in the flank region and the embryo transferred through the uterine wall into the lumen. For single (ipsilateral) transfer of high cost embroys this is still the method of choice because consistently high pregnancy rates can be achieved.

Non-surgical transfer

The embroy is placed in a 0.25 or 0.5 ml semen straw and using the normal insemination gun is deposited into the uterine horn. With this technique care must be taken both to avoid trauma and infection. For contralateral transfer, to induce twin pregnancy in inseminated recipients, this is the current method of choice. Non-surgical transfer is fast and inexpensive and indeed is rapidly gaining popularity even for high cost embroys.

The pregnancy rates achieved following various transfer procedures from a number of studies at this laboratory are summarised in Table 3.

The higher and most consistent pregnancy rates (about 70%) have been achieved following single flank transfer of freshly collected embroys. The overall rate following single non-surgical transfer of similar embroys is about 20% less than that achieved by the Flank method even though pregnancy rates of 60-63% have been achieved in particular experiments. Single ipsilateral non-surgical transfer is not as consistent as similar Flank transfers. The biggest single factor is effecting a high non-surgical pregnancy rate is probably the skill and experience of the operator.

Following non-surgical contralateral transfer (for twinning) overall transferred embryo survival was 62% in pregnant recipient animals.

This represents a twinning rate in pregnant recipients of close to 60%. However, again here the variability of pregnancy rate is greater than with Flank transfers.

Donor-Recipient Synchrony

The recipient animals must be at the same or nearly the same cycle stage as the donor. Highest pregnancy rates are achieved with exact synchrony but a tolerance of $\pm$ 1 day will still give acceptable pregnancy rates. Synchronisation procedures based on wither prostaglandin or short-term progestagen administration are now available and normal pregnancy rates can be obtained following their use.

Potential for using embryo transfer to increase beef production from the dairy herd

There are two possible ways in which embryo transfer might effect an increase in beef output from the dairy herd.

Replacement breeding

By increasing the reproductive rate of a small proportion (about 10%) of the cows it would be possible to breed all the replacements from this small number of cows, leaving the remaining 90% of the herd available for beef crossing. However, we must assume that the dairy herd is at or near the maximum stocking rate and therefore, the herd itself must form the recipient herd.

The effect of various breeding strategies on beef output is shown in Table 4. In all cases a minimal replacement rate of

Table 4 Effect of Embryo transfer (E.T.) on beef output in the dairy herd

1. Replacement Breeding

Breeding strategy	% Pure Friesian	% Beef cross "Small"**	"Large"**
1.	100	-	-
2.	50	20	30
3.	50	5	45
4. +(E.T.)*	50	8	42
5. +(E.T.)*	50	8	42 (Pure beef)

* Top 10% Producing Cows = Donors
E.T. Preg. Rate = 55%

** Small and large refers to use of sire.

Table 5 Effect of Embryo transfer (E.T.) on beef output in the dairy herd

2. "Limited" Twinning

Breeding strategy	% Pure Friesian	% Beef cross "Small"	"Large"	Extra beef Calves
3.	50	5	45	-
3+(E.T.)	50	5	45	+ 23

Table 6. Effect of Embryo transfer (E.T.) on Beef output in the dairy herd

3. "Total" Twinning

Breeding strategy	% Pure Friesian	% Beef cross "Small"	"Large"	Extra beef Calves
6.	95	5	-	40
7.	85	5	10	40

20% is assumed thus requiring 50 Friesian x Friesian calves to be born. Strategy No. 1. does not allow for any beef crossing. Strategy No. 2 allows for half the progeny to be beef cross, with the heifers calving to 'smaller' beef bulls and all replacements born from the mature cows. Strategy No. 3 provides for the calving of most of the heifers to proven dairy bulls thus leaving a greater portion of the mature cows to calve to the larger, faster growing beef bulls. This strategy allows for the greater proportion of beef crossing.

Using Strategy No. 4, one round of superovulation of the top 10 producing cows and the transfer of their embroys to heifer heifers (20) and mature cows (50) allows for 50 Friesian x Friesian calves to be born (including 10 directly from the donor cows). Because the herd itself is used as the recipient herd the net result, however, is not to change the total proportion of beef calves born (50%) but in fact to cause a slight reduction in those calves born from the larger fast growing bulls.

Instead of beef crossing, the mature cows could of course, recieve transplant 'pure' beef embroys thus increasing beef output. However, the likely dystokia problems here would at the present time militate against the use of such a procedure.

The overall effect of using single embryo transfer is unlikely to increase the proportion of beef originating in the dairy herd above what could be achieved using optional breeding strategies.

Twinning

Egg transfers to induce twinning in a proportion of the dairy cows would increase the total calf crop and in this way increase output.

Table 5, outlines the potential effect of induction of twinning in that portion of the herd not normally required for replacement breeding, viz., the late calving and low producing cows. Embryo transfer to the portion of the herd available for beef crossing therefore, would result in 23% of extra calves.

Table 7 Effect of Embryo transfer "Twinning" on the calf crop from the dairy herd.

4. (Composition of Calf Crop)

		Breeding strategy			
		3	3+(E.T.)	6+(E.T.)	7+(E.T.)
	Total calves	100	123	140	140
	Heifers available for breeding	20/25	20/25	20/37	20/33
Calves for beef	FR. x FR.	30	30	75	65
	FR. x Beef (Large)	45	45	-	10
	FR. x Beef (Small)	5	5	5	5
	Transplant Calves (¾ beef)	-	23	40	40
Calves for Beef		80	103	120	120

If the herd is comprising already, cows, bred themselves from proven dairy sires we could assume that there is only a marginal benefit to be gained from selecting particular cows for breeding replacements and in this case we could envisage embryo transfer to the total herd to induce twinning.

The effect of this is shown in Table 6. In breeding Strategy No. 6 all of total mature cows are bred to produce Friesian x Friesian calves and transferred to with a beef embryo. 50% of cows will then also provide a beef calf from the transfer. This strategy would allow all the replacements to come from single calvings and would yield 40 extra calves for beef.

Any further increment of beef output (using a twinning level of 50%) could only come from reducing the number of Friesian inseminations and replacing these with beef insemination. However, the Friesian inseminations can only be reduced by about 10 cows in order to still retain the 20% replacement rate.

The breed composition of the calf crop is set out in Table 7. from selected breeding and transfer strategies. Strategy No. 7 maximises both overall calf crop and the proportion bred from beef sires.

References

Boland, M.P., Kennedy, L.G., Crosby, T.F. & Gordon, I. 1981. Superovulation in the cow using PMSG or HAP. Theriogenology, 15, 1, 110 (Abstr.)

Renard, J.P., Heyman, Y. & Ozil, J.P. 1981. Freezing bovine blastocysts with 1.2 propanediol as cryoprotectant. Theriogenology, 15, (1) 113 (Abstr).

Willadsen, S.M., Polge, C. & Rowson, L.E.A. 1978. *In vitro* storage of cattle embryos. In, Control of Reproduction in the cow. Ed. J.M. Sreenan, Current topics in Vet. Med. Vol. 1, pp. 427-436.

Wilmut, I. & Rowson, L.E.A. 1973. Experiments on the low temperature preservation of cow embryos. Vet. Rec. 92, 686-690.

DISCUSSION ON DR. J. SREENAN'S PAPER

Chairman: Prof. E.P. Cunningham

Thiessen: (a) When you superovulate a cow, do you try to keep her pregnant, or re-implant ? (b) Why is it necessary to maintain original pregnancy to get twinning and is a single implantation always on the same side ?

Sreenan: (a) We superovulate a cow as early as possible in the season, then put her in calf at her next cycle. This will be a little longer than normal, but can be shortened by use of hormones. After flushing, the chances of getting a cow pregnant at the same cycle are low. (b) This is a local effect; the cow ovulates and is inseminated, and pregnancy is maintained on the side of ovulation by means of a signal from the embryo to the corpus luteum. The chances of an ovum implanted into the opposite horn initiating the signal necessary for survival are slim. A single implantation is always to the same horn.

Smidt: Have you any figures on the individual repeatability of superovulation response in repeated superovulation procedures in the same animal; in other words, could superovulation response be regarded as an individual characteristic, and therefore be subject to selection ?

Sreenan: We ourselves have only a small amount of data, but from the literature it appears that up to about three superovulations there is no fall-off in productivity. There is evidence to indicate that this trait is repeatable.

Liboriussen: Could you comment on the possibility of sexing eggs or embryos before implantation.

Sreenan: It is possible to sex the embryo on day 12 or 13. The problem is that these embryos cannot be frozen at that stage (and conversely it is not possible to sex an embryo at day 7 when one can freeze them). At this point in time, using the karyotype technique, efficiency is low, but I am sure another method will be found.

Menissier: What are the costs involved at all stages ?

Sreenan: This is a difficult question to answer. In the non-surgical transfer, a high proportion of cost is travel expenses; otherwise the time involved is similar to that for artificial insemination. The main cost is the cost of embryo production. In the case of twinning, it is possible to reduce costs considerably; for example, the cost of production of embryos in a feedlot system with heifers would be governed by economic efficiency of the feedlot enterprise.

Zervas: Can you comment on the possibility of cloning of cattle eggs ?

Sreenan: We are not yet at the stage of discussing the practical applications of cloning in cattle breeding. Perhaps Professor Gordon of University College Dublin, would care to comment on this topic.

Gordon: Cloning has been going on in the frog world since 1952, by taking the nuclei from somatic cells and putting them into frog eggs, and getting an individual which is a copy of the individual which donated the somatic nucleus. Over the past two to three years there have been successful nuclear transplants in mammals, that is in mice. Two workers in the USA and Illmensee in Geneva have successfully cloned mice. They took somatic nuclei from the ectoderm of early developing mouse embryos, and introduced them into previously fertilised mouse eggs where male and female nuclear material was removed. There is no reason why these techniques and procedures cannot be applied in cattle. It involved micro-surgical techniques, and is referred to as 'reproductive engineering.' I feel sure that advances in this area will come in cattle.

Smidt: (a) Do you employ any selection procedures among the donor cows with reference to their reproductive status ?
(b) Do you think that farmers will accept twinning on a large scale, especially in dairy cattle, in view of the disadvantages associated with it, for example lower birth weight, higher mortality, retained placentas ?

Sreenan: (a) I have been talking about beef cross embryos from heifers (Hereford X Friesian) so selection on previous reproductive rate is not an issue. In relation to dairy females it will become more important. If a number of lactations are needed to assess the breeding value of the females, then these will be expected to be poorer donors. It is generally also found that Friesians are poorer donors, in other words, there are breed differences.

(b) Yes, historically there is adverse reaction to twins. In the past, twin-bearing cows have been managed as if they were carrying only singles. Now we have a blood test which identifies twin-bearing cows, and this will allow better management during pregnancy and result in lower placental retention.

Neimann-Sorensen: Can we use cull cows as recipients in order to increase the output of beef from dairy herds ?

Sreenan: It is possible, but not very effective. These are old cows and are producing small numbers of embryos. Also, the reproductive state of the recipient is very important.

Diskin: Micro-manipulation offers the opportunity for splitting embryos and so reducing the number of donors required to produce a given number of embryos.

GENERAL DISCUSSION

Oostendorp: The overall economic consequences of using Holsteins, that is both the dairy and beef production aspects, needs to be considered.

Oldenbroek: The most negative trait in Holsteins is fat percentage, and we can select readily to improve this. However, we also need to incorporate the negative aspects of meat production.

Cunningham: I seem to recall that Dutch figures indicated that the first cross (Holstein X Dutch Friesian) is more productive than Dutch Friesian, whereas purebred Holstein is economically no better overall than Dutch Friesian, where both milk and meat are taken into account. Is that correct ?

Oldenbroek: Yes.

Langholz: As a general conclusion from discussions on breeding strategies in dual-purpose cattle, it should be realised that overemphasis on dairy traits is, to some extent, due to the fact that recording of beef traits is not consistent. Perhaps the approach of Colleau, which utilises the market value of 10-day old calves in sire evaluation, is worth considering.

Cunningham: In most countries farmers are, in fact, using Holsteins or Brown Swiss against the advice of their Governments.

Jansen: There is no reason not to incorporate the useful genes of such breeds, but the undesirable beef genes should also be eliminated by performance testing. Imported breeds should be compared with the traditional breeds in a fair manner.

Neimann-Sorensen: We recognise both negative and positive effects of Holsteins and the right approach is to counter the negative effects by selection. I am reluctant to accept that there is controversy between AI organisations and farmers.

Cunningham: Has it not been the case that breeding organisations in some countries have obstructed the introduction of exotic semen (through higher fees, etc.) ?

Southgate: To some extent in the UK this is true, where there is a tradition of awkwardness in relation to the use of AI.

Oostendorp: Farmers were not aware of the beef quality of their cows, they were just paid more for calves sired by beef bulls.

Cunningham: It is clear that there is a unanimous feeling that there is a deficiency of balanced data on new breeds. There is need for (a) more effective screening in selection of animals in the countries of origin, and (b) more effective screening on introduction into AI in the host country.

Bittante: The majority of trials evaluating the effect on milk and meat production of the diffusion of genes from the American highly specialised milk breeds (Holstein Friesian, Canadian Friesian, Red Holstein, and Brown Swiss) are based on the comparison between the performance of the offspring sired by bulls of these breeds mated to cows of the European breeds and their contemporary European purebreds. We cannot therefore generalise the results obtained because they are partially due to heterotic effects. In future it is important to try to disentangle the heterotic and additive genetic effects in order to avoid possible over-estimation of the genetic merit of these breeds.

Fewson: In their paper, Colleau and Lienard showed that breeds perform differently in different environments. We should therefore consider not only purebreds, but also breeding systems to take into account the different systems of production.

Langholz: I want to support that comment of Fewson; I agree that breeding policies should be more orientated to system approaches, in what I call regional breeding and production planning. For the average farm conditions we calculate that 85% of genetic progress should come from dairy traits and about

15% from beef traits. However, we still have a stable population of the more beefy type dual-purpose breeds, Fleckvieh and Red and White. Perhaps we should also think of inter-European strategies; for example, I could imagine that commercial cross-breeding with specialised beef breeds could be applied in the Western part of Lower Saxony with Piedmontese, if we could arrange for a female crossbred calf market for the Dutch veal production industry.

Cunningham: This is a very important point. Among the 25 million dairy cows in Europe, some breeds have not been touched by this shift to the extreme. For example, Normande, Fleckvieh and the MRY. Should we learn from our experience with the Holstein Friesian and protect these dual-purpose breeds ?

Langholz: The important thing is to document the information on beef aspects (through calf market sources) and use this in an index.

Noble: Dual-purpose breeds have not been able to survive in the UK. The Shorthorn and Red Poll breeds have almost disappeared and the attempt to revive the Shorthorn has relied heavily on the Red and White Friesian. The Welsh Black, Lincoln Red and South Devon have all changed from dual-purpose breeds to beef breeds.

CHAIRMAN'S SUMMARY AND CONCLUDING REMARKS

Chairman : Prof. E.P. Cunningham

In this Seminar, we have systematically reviewed the evidence on the extent to which North American Holstein and Brown Swiss genes are being incorporated into the different European dairy cattle populations, and the effect that this is having on the beef potential of these populations. It is clear from the evidence presented that, as a result of trends already established, most European Friesian populations are likely to become considerably more than 50% North American in constitution. The studies reported suggest that Holstein crosses are from 5-10% less valuable for beef production than the existing Friesian strains, though carcass dissection studies indicate a smaller difference in value, while actual market prices for animals show even bigger discrimination.

Despite the surplus of milk and near deficit of beef in the European market, there is little evidence that the pressures which have led to this swing toward North American strains will change in the immediate future. Indeed, there is evidence in many countries, for example that presented by Colleau on the situation in the Normande breed, that similar trends may be expected in the more firmly dual purpose populations such as the MRI, the Flechvieh and the Normande.

It is an anomaly worth some reflection in Brussels and elsewhere that we appear to be in the process of converting 25 million dairy cows in the EEC to North American genotypes in order to better utilise the 30 m tonnes of imported feed grain, also largely North American, in a system of intensive dairy production which is causing great difficulties in the market, is devaluing the basic ruminant function of the cow, and is leading to a deterioration in the beef producing capacity of our industry.

What can, or should, be done to preserve the beef potential of Europe's dairy breeds ? Several of the papers presented have indicated that the economic returns from increased testing and selection for aspects of beef merit in dual purpose populations are very high. The contribution of Fewson indicated that there is a clear difference in the relative weightings of milk and beef traits when viewed from the national standpoint and from the standpoint of the individual farmer, with the public interest requiring a much heavier weighting on beef. These two considerations lead to the conclusion that much greater effort in terms of testing facilities and selection intensities should be applied to beef traits. This is particularly important in the screening of genetic source material from North America.

The other breeding strategy which can lead to an improvement in the quantity and quality of beef from the dairy herd is the use of specialised beef breeds for crossing. The papers presented on this topic have confirmed the technical advantages of the crosses from certain breeds in terms of growth and carcass merit. They have also confirmed the disadvantages, primarily on the question of higher incidence of calving difficulties. However, it is clear that while beef crossing continues at a high level in Ireland, Britain and to a lesser extent France, it has declined in the Danish Jersey population, and has not made any significant contribution in most other populations. The reasons for this situation have been discussed and the conclusions are that the present pattern is likely to continue.

I believe this Seminar has assembled a very comprehensive, or authoritative, and up-to-date survey of the information concerning beef production from the European dairy herd. It is clear that the current evolution of that population is not favourable to improve beef production. Furthermore, it is clear that this evolution represents the collective response of European dairy farmers to the economic opportunities and pressures which they face. In particular, two price ratios appear to play a major part : milk/meat, and milk/concentrate feed. Altering the economic order to which producers must respond is a matter which concerns a great deal more than genetic suitability of our dairy cattle for beef production.

LIST OF PARTICIPANTS

BELGIUM

Mr. Ch.V. Boucqué	Rijksstation voor Veevoeding, Scheldeweg 68, B 9231 Melle-Gontrode
Mr. L.O. Fiems	"
Prof. Dr. J.M. Bienfait	Faculté de Médicine Vétérinaire, Rue des Vétérinaires 45, Bruxelles

DENMARK

Dr. B. Bech Andersen	National Institute of Animal Science, Rolighedsvej 25, DK-1958, København V
Dr. A. Neimann-Sorensen	"
Dr. T. Liboriussen	"

FRANCE

Dr. J.J. Colleau	Station de Génétique quantitative et appliquée, Centre National de Recherches Zootechniques, I.N.R.A. 78350 Jouy-en-Josas
Dr. F. Menissier	"
Mr. J. Sornay	I.T.E.B./C.E.R.C.V. Route d'Epinay, 14130 Villers-Bocage
Dr. J.L. Foulley	Station de Génétique quantitative et appliquée, Centre National de Recherches Zootechniques, I.N.R.A. 78350 Jouy-en-Josas

GERMANY

Prof. Dr. H.J. Langholz — Institut für Tierzücht und Haustiergenetik,
Albrecht Thaerweg 1,
3400 Göttingen

Prof. Dr. Smidt — Institüt für Tierzucht und Tierverhalten (GAC),
Mariensee,
3057 Neustadt 1

Prof. Dr. D. Fewson — Institüt für Tierzucht und Tierzüchtung,
Garbenstrasse 17,
7000 Stuttgart 70

GREECE

Prof. N.P. Zervas — Department of Animal Husbandry,
Aristotelian University of Thessaloniki,
Thessaloniki

IRELAND

Prof. E.P. Cunningham — The Agricultural Institute,
(An Foras Taluntais),
19 Sandymount Avenue,
Dublin 4

Dr. J. Flanagan — Department of Agriculture,
Setanta Centre,
Nassau Street,
Dublin 2

Dr. J.M. Srennan — The Agricultural Institute,
Western Research Centre,
Belclare,
Tuam, Co. Galway

Dr. T.J. Teehan — Department of Agriculture,
Setanta Centre,
Nassau Street,
Dublin 2

Prof. I. Gordon — Department of Agriculture,
University College,
Lyons,
Newcastle, Co. Dublin

Dr. G.J. More O'Ferrall — The Agricultural Institute,
Dunsinea,
Castleknock,
Co. Dublin

ITALY

Dr. A. Romita	Istituto Sperimentale per la Zootecnica, Via Salaria 31, 00016 Monterotondo Scale, Roma
Dr. G. Bittante	Istituto di Zootecnica, Università di Padova, Via Gradenigo 6, 35100 Padova
Dr. A.L. Catalano	Istituto di Zootecnica Alimentazione e Nutrizione, Facolta di Medicina Veterinaria, Università di Parma, Cornocchio, 43100 Parma
Dr. S. Gigli	Istituto Sperimentale per la Zootecnica, Via Salaria 31, 00016 Monterotondo Scale, Roma

LUXEMBOURG

Mr. E. Wagner	Agricultural Technical Services, B.P. 1904, 16 Route d'Esch

NETHERLANDS

Ir. J.K. Oldenbroek	Instituut voor Veeteeltkundig Onderzoek "Schoonoord" P.O. Box 501, 3700 AM Zeist
Ir. J. Jansen	Agricultural University of Wageningen, Landbouwhogschool, Marijkeweg 40, 6709 PG Wageningen
Ir. D. Oostendorp	Proefstation voor de Rundveehouderij, Runderweg 6, 8219 PK Lelystad

POLAND

Dr. Z. Reklewski (not present)	Instytut Genetyki i Hokowli, Zwierzat Pan, Jastrgebiec, 05-551 Mrokow

UNITED KINGDOM

Dr. R.B. Thiessen	Animal Breeding Research Organisation, King's Buildings, West Mains Road, Edinburgh EH9 3JA, Scotland
Mr. J.R. Noble	Regional Livestock Husbandry Adviser, Block B, Government Buildings, Brooklands Avenue, Cambridge CB2 2DR, England
Mr. J.R. Southgate	Meat and Livestock Commission, P.O. Box 44, Queensway House, Bletchley, MK2 2EF, Bucks. England

EEC COMMISSION

Mr. J. Connell	CEC Direction Général de l'Agriculture, DG VI, 200 Rue de la Loi, B-1049 Brussels
Mr. G. Van Snick	CEC Direction Général de l'Agriculture, Div B VI, 200 Rue de la Loi, B-1049 Brussels

FSC
www.fsc.org
MIX
Papier | Fördert gute Waldnutzung
FSC® C083411

Kolibri 360 GmbH
Ferdinand-Jühlke-Straße 7
99095 Erfurt, Deutschland
produktsicherheit@kolibri360.de